# Analyzing International Environmental Regimes

**Global Environmental Accord: Strategies for Sustainability and Institutional Innovation**
Nazli Choucri, series editor

Nazli Choucri, editor, *Global Accord: Environmental Challenges and International Responses*

Peter M. Haas, Robert O. Keohane, and Marc A. Levy, editors, *Institutions for the Earth: Sources of Effective International Environmental Protection*

Ronald B. Mitchell, *Intentional Oil Pollution at Sea: Environmental Policy and Treaty Compliance*

Robert O. Keohane and Marc A. Levy, editors, *Institutions for Environmental Aid: Pitfalls and Promise*

Oran R. Young, editor, *Global Governance: Drawing Insights from the Environmental Experience*

Jonathan A. Fox and L. David Brown, editors, *The Struggle for Accountability: The World Bank, NGOS, and Grassroots Movements*

David G. Victor, Kal Raustiala, and Eugene B. Skolnikoff, editors, *The Implementation and Effectiveness of International Environmental Commitments: Theory and Practice*

Mostafa K. Tolba, with Iwona Rummel-Bulska, *Global Environmental Diplomacy: Negotiating Environmental Agreements for the World, 1973–1992*

Karen T. Litfin, editor, *The Greening of Sovereignty in World Politics*

Edith Brown Weiss and Harold K. Jacobson, editors, *Engaging Countries: Strengthening Compliance with International Environmental Accords*

Oran R. Young, editor, *The Effectiveness of International Environmental Regimes: Causal Connections and Behavioral Mechanisms*

Ronie Garcia-Johnson, *Exporting Environmentalism: U.S. Multinational Chemical Corporations in Brazil and Mexico*

Lasse Ringius, *Radioactive Waste Disposal at Sea: Public Ideas, Transnational Policy Entrepreneurs, and Environmental Regimes*

Robert G. Darst, *Smokestack Diplomacy: Cooperation and Conflict in East-West Environmental Politics*

Urs Luterbacher and Detlef F. Sprinz, editors, *International Relations and Global Climate Change*

Edward L. Miles, Arild Underdal, Steinar Andresen, Jørgen Wettestad, Jon Birger Skjærseth, and Elaine M. Carlin, *Environmental Regime Effectiveness: Confronting Theory with Evidence*

Erika Weinthal, *State Making and Environmental Cooperation: Linking Domestic and International Politics in Central Asia*

*(continued at back of book)*

# Analyzing International Environmental Regimes

## From Case Study to Database

Helmut Breitmeier, Oran R. Young, and Michael Zürn

The MIT Press
Cambridge, Massachusetts
London, England

MIT Press books may be purchased at special quantity discounts for business or sales promotional use. For information, please e-mail special_sales@mitpress.mit.edu or write to Special Sales Department, The MIT Press, 55 Hayward Street, Cambridge, MA 02142.

This book was set in Sabon on 3B2 by Asco Typesetters, Hong Kong.
Printed and bound in the United States of America on recycled paper.

Library of Congress Cataloging-in-Publication Data

Breitmeier, Helmut.
Analyzing international environmental regimes: from case study to database / Helmut Breitmeier, Oran R. Young, and Michael Zürn.
p. cm. — (Global environmental accord)
Includes bibliographical references and index.
ISBN 0-262-02602-3; 978-0-262-02602-4 (hc : alk. paper) —
ISBN 0-262-52461-9; 978-0-52461-2 (pbk. : alk. paper)
1. International regimes database. 2. Environmental sciences—Information services. 3. Environmental sciences—Databases. 4. Environmental management—International cooperation. I. Young, Oran R. II. Zürn, Michael. III. Title. IV. Global environmental accords.
GE32.B74 2006
363.7′0526—dc22 2006041964

10 9 8 7 6 5 4 3 2 1

# Contents

# Series Foreword

A new recognition of profound interconnections between social and natural systems is challenging conventional constructs and the policy predispositions informed by them. Our current intellectual challenge is to develop the analytical and theoretical underpinnings of an understanding of the relationship between the social and the natural systems. Our policy challenge is to identify and implement effective decision-making approaches to managing the global environment.

The series Global Environmental Accord: Strategies for Sustainability and Institutional Innovation adopts an integrated perspective on national, international, cross-border, and cross-jurisdictional problems, priorities, and purposes. It examines the sources and the consequences of social transactions as these relate to environmental conditions and concerns. Our goal is to make a contribution to both intellectual and policy endeavors.

Nazli Choucri

# Preface

# The Long Road to Cumulative Knowledge

The idea of constructing a regimes database arose during the November 1991 Regimes Summit. That workshop (supported in part by Ford Foundation grant 890-0042-4) brought together leaders of four major projects—the Oslo/Seattle Project, the Tübingen Project, the Dartmouth Project, and the Harvard Project—who sought to "arrive at common ground among a number of projects dealing with international regimes in the interests of promoting the development of cumulative knowledge about the origins, operations, and outcomes of various forms of institutionalized international cooperation" (Regimes Summit 1991, 1). The Minary Center, owned and operated by Dartmouth College, provided an ideal setting for this gathering.

Participants in the 1991 workshop understood the magnitude of this undertaking; agreement would have to be reached on the design of a protocol that could be used in constructing a regimes database accessible electronically to those engaged in formulating and testing ideas about international regimes. Understandably, therefore, it took some time before we could begin to implement this action plan. The opportunity to move forward came in 1994 when the International Institute for Applied Systems Analysis (IIASA) initiated a three-year project on the Implementation and Effectiveness of International Environmental Commitments (IEC) (Victor, Raustiala, and Skolnikoff 1998). The creation of an International Regimes Database (IRD) emerged as one of a number of lines of inquiry included in this project. Marc A. Levy, Oran R. Young, and Michael Zürn undertook to provide scientific leadership for the effort to create the IRD. Early on, we realized that progress toward this goal would require the involvement of a full-time project manager located at

IIASA. This led to the addition of Helmut Breitmeier to the IRD team. He spent two and a half years in residence at IIASA as the database manager during the period 1995–1997; he has remained a central player in the development of the IRD throughout its history.

The IIASA phase of the project focused on the development of what we now know as the IRD Data Protocol (Breitmeier et al. 1996a). The protocol provides definitions of key terms to be used by all participants in the project and identifies all the variables included in the database. The protocol constantly threatened to explode into an unmanageable instrument; its final version is relatively long and complex, despite a concerted effort to control unlimited expansion. The protocol emerged from an ongoing dialogue between the members of the database team and many advisors, including participants in a series of trial runs. Numerous scientists working both at IIASA and at other research institutes located for the most part in Europe or North America provided advice regarding the design of the data protocol.

The IEC project and other projects at IIASA provided a stimulating intellectual environment for our work; we have maintained intellectual cooperation with many of these colleagues following the end of the IEC project. We are particularly grateful to those who participated in trial runs of the database protocol at IIASA. The discussions we had with these experts during the course of the trial runs allowed us to improve the protocol substantially and to make it more user friendly. The results are noticeable particularly in the development of the idea of the precoding agreement and in the guidelines governing the development of case structures necessary for the coding of individual regimes.

When the IIASA project came to a close in 1997, the data protocol for the IRD was essentially complete. But the database itself was an empty shell devoid of data. The U.S. National Science Foundation (NSF) provided modest funding that helped keep the project going during a transitional period (grant 9631659). The German-American Academic Council (GAAC) awarded a travel grant that allowed us to hold research meetings among project members in the years following the end of IIASA's IEC project. The German Science Foundation (DFG) ultimately came through with substantial support needed to move forward from the design stage of the IIASA years to the operational stage featuring the coding of individual cases, the reconciliation of differences between coders, the computerization of all data resulting from the coding effort, and the construction of data tables accessible in MS Access (grants WO 400/5-1 and WO 400/5-2).

During this stage, lasting from 1997 through 2002, the IRD's base of operations moved to Darmstadt University in Germany. We are especially grateful to the Department of International Relations at the Institute for Political Science in Darmstadt for the support we received during this period. We acknowledge as well the essential role during this stage of the hard-working student assistants who checked the data the coding experts delivered and entered these data in the IRD. Members of the Department of Mathematics at Darmstadt solved a number of problems pertaining to the management and evaluation of data.

During this period, Helmut Breitmeier continued to devote most of his professional time and energy to the project. Marc Levy took on a new position at CIESIN that has made it impossible for him to play an active role in the project in recent years. Oran Young and Michael Zürn maintained their roles as key members of the IRD management team. This explains the authorship of this book.

We learned during the project that the effort required for experts to code individual cases is substantial. The honoraria we were able to provide to coders did not compensate them adequately for the time and energy required to code an individual case. Accordingly, we are especially grateful to our expert coders and pay tribute to their commitment to the creation of this database. We admire their legal, political, or scientific expertise; the IRD could not have come on stream without the invaluable input provided by case-study experts. Our thanks, then, go to the following coders: Matthew Auer, James S. Beckett, Thomas Bernauer, Pamela Chasek, René Coenen, Elisabeth Corell, Debbie Davenport, Leonard B. Dworsky, Peter Ehlers, Christel Elvestad, David S. Favre, Bob Friedheim, Fred Gale, Ray Gambell, Andy Garner, Brian Hallman, Gudrun Henne, Richard Herr, Geir Honneland, James Joseph, Christopher Joyner, Jonathan Krueger, Jack Manno, Frank Marty, Britta Meinke, Radu Mihnea, Ronald Mitchell, Ilia Natchkov, Sebastian Oberthür, Kate O'Neill, Edward A. Parson, Dwight Peck, Gerard Peet, M. J. Peterson, Kal Raustiala, Lasse Ringius, Peter Sand, Sibel Sezer, Clare Shine,

Jon Birger Skjaerseth, David Victor, Virginia Walsh, Jacob Werksman, Jørgen Wettestad, Koos Wieriks, Andrea Williams, and Bernhard Zangle. Information about the case each coder worked on is readily available in the database itself.

Thanks should also go to the Centre for Interdisciplinary Studies in Technology (ZIT) of Darmstadt University of Technology, which awarded a grant for the development of a TOSCANA system. This system provides a graphical tool that allows users to study relationships between variables included in the IRD. Appendix B describes and illustrates the use of the TOSCANA system.

Our intention at this juncture is to make the IRD available to interested members of the research community; we hope that the results will help overcome some of the obstacles to the development of cumulative knowledge regarding the formation, operation, and effectiveness of international regimes. To this end, we are including a CD-ROM containing the database itself with each copy of the book. Thus, this book is not only scientifically significant in its own right; it also marks a crucial turning point in the life of the project. In the substantive chapters of the book, we present some early examples of the types of analysis made possible through the use of the IRD. With the publication of this book, we declare the IRD open for use on the part of other members of the research community. This is a kind of debut or coming out in which a project whose design and developmental stages have lasted longer than any of us could have anticipated is now ready finally for use on the part of others as a public good.

As is often the case with labor-intensive projects, we might well have thought twice about embarking on the creation of the IRD if we had been fully aware in 1994 of what it would take to reach the current stage in the life of the project. Yet we believe that the IRD has much to offer to members of the research community concerned with international regimes; it represents a qualitative advance over past research on international regimes. We look forward to the results produced by other users who see in the IRD a useful research tool.

We are grateful to all who have assisted us in our journey along the long road to cumulative knowledge, and for the general support provided by the Bren School of Environmental Science and Management,

University of California, Santa Barbara; Dartmouth College; the Institute of Political Science, Darmstadt University of Technology; the Social Science Research Center, Berlin; and the Institute for Intercultural and International Studies at the University of Bremen.

Of course, we accept responsibility for any errors of commission or omission remaining in the book. But the book itself could not have come into existence without the help of many colleagues and friends.

# 1

# Regime Theory: Achievements and Challenges

## 1.1 Introduction

Launched in the 1970s as a reaction to the formalism of mainstream work on international organizations, regime theory has grown into a major movement among analysts focusing on international relations and, more generally, world affairs. The regime-theoretic approach has taken root with particular vigor in the subfields of international political economy and international environmental politics. Thus, we now have a voluminous literature on regimes dealing with trade and monetary concerns and on regimes addressing environmental concerns ranging from bilateral issues (e.g., the protection of the Great Lakes in North America) to global issues (e.g., the protection of the stratospheric ozone layer). But there is also a good deal of interest in regime-theoretic work pertaining to other issue areas, such as arms control and human rights.

Naturally, regime theory has its detractors as well as its supporters. Critics have taken regime theory to task for a variety of alleged shortcomings, including a lack of concern for power politics and a failure to understand that institutions at the international level are really epiphenomena distracting analysts who should be concentrating on the real driving forces (e.g., power in the material sense) in this realm. Nonetheless, it is undeniable that regime theory has grown over the last twenty-five years into a productive and robust stream of analysis whose influence now extends throughout the world.

What accounts for the success of regime theory? Part of its appeal surely lies in the theory's perspective on governance in a world in which the demand for governance is great and growing but the familiar mechanisms for supplying governance (e.g., the various elements of the UN System) are notoriously weak. Regimes offer no panacea when it comes to meeting the demand for governance. Some efforts to create regimes end in failure; many issue-specific regimes perform poorly. Yet the general idea of governance without government is highly attractive (Rosenau and Czempiel 1992). At a minimum, it suggests that efforts to respond to the demand for governance at the international level do not constitute a lost cause. Beyond this, regime theory provides an umbrella that is broad enough to encompass a wide range of epistemological preferences and theoretical perspectives. There is room for those who focus on the roles of power, interests, or knowledge to interact constructively with one another in analyses of the formation and operation of specific regimes (Hasenclever, Mayer, and Rittberger 1997). Regime theory has no difficulty accommodating the contributions of constructivists as well as an array of analysts who regard themselves as positivists (Rittberger 1993; Young 1999b).

All this is gratifying for those of us who have worked on issues relating to international regimes for a long time. But it will come as no surprise that regime theory is plagued with a number of problems that have proven difficult to overcome. There are, to begin with, problems of delimitation. Although most would agree on the identity of a core group of regimes, there are ragged edges that make it hard to determine the exact boundaries of the universe of cases in this realm. Ongoing difficulties hamper efforts to demonstrate the causal significance of regimes. At the most basic level, this problem arises from the fact that measures of regime effectiveness typically subsume—implicitly if not explicitly—a causal judgment. But it is also difficult to separate the impact of regimes from the influence of other factors operating at the same time in such a way as to arrive at persuasive conclusions concerning the proportion of the variance in collective outcomes that can be attributed to the operation of institutional arrangements. So far, our efforts to develop cumulative knowledge about international regimes have borne only a limited amount of fruit. Although analysts have proposed a wide range of hypotheses about the formation and operation of regimes, it is generally easy to provide counterexamples that falsify these hypotheses.

What can we do to overcome these problems and, in the process, take the next step in realizing the potential of regime theory? This book constitutes the first major report on a project initiated over a decade ago to address these problems, facilitating in the process our efforts to deal with a range of key issues on a sustained, comparative, and empirically grounded basis. An international workshop held in 1991 provided the intellectual foundation for this project, known as the International Regimes Database or IRD. Support from the International Institute for Applied Systems Analysis (IIASA) made it possible to take the first concrete steps toward building the IRD in 1994. Subsequent support from American and German funders allowed the project to continue in the aftermath of IIASA support.

The goal of this book is to present the IRD to the community interested in the analysis of international regimes, develop a few applications that illustrate the range and types of analysis that the IRD makes possible, and discuss opportunities for interested members of the community to make use of the IRD to illuminate a variety of other issues.

## 1.2 The Research Agenda

International regimes are social institutions created to respond to the demand for governance relating to specific issues arising in a social setting that is anarchical in the sense that it lacks a centralized public authority or a government in the ordinary meaning of the term. Arrangements of this sort have long been a part of the sociopolitical landscape at the international level. But there is no doubt that the ranks of international regimes have grown rapidly since the close of World War II. The result is a proliferation of arrangements addressing a wide range of concerns from functional issues that are global in scope (e.g., the promotion of free trade and the protection of endangered species) to spatially delimited functional issues (e.g., the control of pollution in the North Sea and the governance of human activities in Antarctica).

This perspective has several implications for studies of regimes that provide structure to the resultant theory. Treating regimes as social institutions makes it natural to link regime theory to the "new institutionalism" spreading throughout the social sciences in recent years and to draw on this movement as an important source of intellectual capital (March and Olsen 1989; North 1990; Rutherford 1994; Scott 1995; Young 1994). Among other things, this line of thought emphasizes the importance of differentiating between regimes construed as sets of rights, rules, and decision-making procedures that give rise to social practices on the one hand and organizations treated as material entities that have offices, personnel, budgets, and so forth on the other. Put simply, regimes provide the rules of the game; organizations typically emerge as actors pursuing their objectives under the terms of these rules.

Equally important is the fact that regimes are ordinarily problem driven. The incentive to create specific regimes arises in connection with efforts to address more or less well-defined problems (e.g., how to stabilize international monetary relations, how to protect the earth's climate system); the willingness to take them seriously depends on an acknowledgment of the importance of the problems in question. As these comments suggest, there are cases in which important players disagree about the nature or significance of the problem as well as cases in which the views of key players regarding the character of the problem evolve or change over time. There remain serious differences, for instance, regarding the nature of climate change as a problem requiring the creation of an international regime as a matter of priority. Whereas some regard the problem of climate change as a matter of controlling concentrations of greenhouse gases in the earth's atmosphere, others approach the problem as a matter of decarbonizing industrial systems. Similarly, many parties now see the international regime dealing with whales and whaling, which began life as an arrangement dedicated to conservation and the achievement of sustainable yields, as a regime whose mission centers on the preservation of individual animals in contrast to the achievement of sustainable harvests from identifiable stocks.

Regimes dealing with specific issues are also embedded institutions in the sense that they operate within the confines of an overarching society. Traditionally, we have assumed that it is broadly accurate to describe this setting as a society of states (Bull 1977), a premise that implies that the members of individual regimes will normally be states and that the rules associated with the idea of sovereignty (e.g., individual states cannot be bound without their consent, outsiders should not intervene in matters within the domestic jurisdiction of states) will apply to the operation of regimes.

Still, there is nothing sacred about this assumption (Keene 2002). The emergence of various nonstate actors as driving forces at the international or transnational level and the growth of what many observers now describe as global civil society are clearly important trends in this realm (Keane 2003). Commentators differ on the extent to which we should now treat various nonstate actors as players in their own right in issue-specific regimes and whether we should be thinking about the growth of regimes in which states do not figure at all as important actors (Cutler, Haufler, and Porter 1999; Khagram, Riker, and Sikkink 2002). These issues will undoubtedly warrant serious consideration during the next stage of regime analysis. But one thing is clear already: issue-specific regimes normally operate in social settings in which there are overarching institutional arrangements that determine the identity of the major actors and configure the deep structure of the broader social environment. This does not mean that it is irrelevant to look at regimes as constitutive arrangements regarding specific issue areas. The Antarctic Treaty System, for example, has a specific criterion dealing with membership in the system. But it does make it important to understand that issue-specific regimes are embedded institutions (Ruggie 1983).

It is easy to understand why many of those studying international regimes have chosen to concentrate first on regime formation or the processes through which specific regimes come into existence. Not only was the proliferation of issue-specific regimes a striking development in the postwar years, but this development also seemed to constitute an anomaly from the perspective of the neorealist thinking then dominant in the field of international relations. Of course, realists can simply assert, as Susan Strange famously did in the early days of regime theory, that regimes in international society are merely epiphenomena, coming into existence as a result of the exercise of power and changing or even vanishing when the political bargains underlying them dissolve or erode (Strange 1983). Yet a wide range of analysts regard this simple interpretation as unsatisfactory. Most practitioners pay a great deal of attention to the roles that institutional arrangements play in international society, a mode of thinking that seems hard to dismiss as nothing more than a manifestation of false consciousness (Chayes and Chayes 1995). What is more, the fact that the new institutionalism has highlighted the roles that institutions play in a wide range of social settings has increased the credibility of the arguments of those who claim that institutional arrangements can account for at least some of the variance in the collective outcomes occurring in international society.

Among those who pay attention to regime formation, a number of distinct, albeit related, questions have come to the fore (Haggard and Simmons 1987; Levy, Young, and Zürn 1995; Hasenclever, Mayer, and Rittberger 1997). We want to know why regimes form in response to some problems but not in response to others, why some regimes form relatively quickly whereas others take years of hard bargaining to see the light of day, and why individual regimes take strikingly different forms (e.g., framework/protocol arrangements versus more comprehensive arrangements) and rely on distinct approaches to problem solving. Underlying these concerns are several generic issues that the study of regime formation shares with studies of public choice in a wide range of social settings. We want to understand the processes through which issues or problems find their way onto the international agenda, are framed for consideration as matters of public policy, and move to a high enough position on the agenda to trigger focused efforts to create an appropriate institutional arrangement. We also want to understand the processes through which interested parties reach agreement on the terms of conventions, treaties, or declarations setting forth the provisions of specific regimes. Is it accurate to describe these processes as a kind of institutional bargaining? If so, does this form of bargaining differ in significant ways from the more familiar processes of legislative bargaining occurring in domestic settings (Young 1994)? And finally, there is the issue of accounting for specific features of the regimes that arise to deal with various problems. Can we explain, for instance, differences in the nature of the decision rules negotiators select in the process of forming specific regimes?

This is a large and important research agenda. Still, a concentration on processes of regime formation is obviously insufficient to produce a convincing case for the allocation of time and energy to the development of regime theory. If regimes should turn out to be largely epiphenomena, the launching of a program of studies designed to improve our understanding of these arrangements would constitute a poor allocation of resources. This is the clear implication of the position adopted by analysts like John Mearsheimer who speak of the "false promise" of international institutions (Mearsheimer 1994–1995). It is no accident, therefore, that research on the effectiveness or the success of issue-specific regimes developed rapidly into a second major pillar of regime theory (Young 2002b). Studies of the effectiveness of regimes must come to terms with several major methodological challenges. So long as effectiveness as such is taken as the dependent variable, efforts to measure this variable must find ways to cope with the fact that this variable has an element of causality embedded in it. Thus, it is pointless to make claims about the effects of regimes, unless we have some confidence that these effects really are consequences of the existence and operation of the relevant governance systems. One way to address this problem is to direct attention to outputs, outcomes, and impacts, where outputs are regulatory and organizational efforts occurring in the wake of regime formation, outcomes are behavioral changes following the creation of a regime, and impacts are changes in the status of the original problem that occur in the aftermath of regime formation.

Assuming that we can operationalize these measures of change in some relatively straightforward manner, it then makes sense to pose a range of questions regarding the roles that regimes play in bringing about these changes. In general, it is fair to say that the shorter the causal chain linking a regime and its effects, the easier it is to demonstrate causality but the less important the results will be in terms of explaining major occurrences in world affairs. Thus, it is easier to draw inferences about outputs than about impacts. Yet it is the impacts of regimes on specific problems that ultimately account for our interest in these arrangements.

Analyses of the effectiveness of regimes must also struggle with the fact that institutions normally interact with a variety of other drivers of collective outcomes in human societies (Young 2002a). The significance of this observation depends on the length and complexity of the causal chains linking institutions and their effects. If individual member states promulgate regulations designed explicitly to implement the terms of an international agreement creating a new regime, it is reasonable to treat this action as a consequence of regime formation. If there is improvement in an environmental problem (e.g., a slowing of the loss of biological diversity) over a period of time following the creation of a regime designed to deal with the problem, on the other hand, we may conclude that any of a number of other economic, political, and social drivers—not to mention biophysical factors—played a role in producing this change. Ultimately, we need to think in terms of the proportion of the variance in a variety of dependent variables that can be attributed persuasively to the operation of regimes. We should regard it as perfectly natural to observe considerable variation in these terms both from one regime to another and within individual regimes over time. In other words, debates about the effectiveness of regimes are seldom likely to result in simple yes or no conclusions.

With the passage of time, a number of other concerns about international regimes have come to supplement studies of regime formation and effectiveness. It is apparent even to the most casual observer that individual regimes—like all social institutions—change continually; they do not remain fixed in time following their initial formation. As a result, efforts to identify patterns of institutional change and to understand the forces giving rise to these patterns have become important concerns (Young 1999b). As issue-specific regimes have proliferated in international society, moreover, it is easy to see that individual arrangements will interact with one another with increasing frequency. This phenomenon has been brought to our attention most forcefully in the so-called trade-environment debate, which addresses a range of interactions between trade regimes and environmental regimes that contain provisions pertaining to trade (Sampson and Chambers 2002). But this is merely the tip of the iceberg in an area destined to become a major focus of interest among students of international institutions (Young et al. 1999). Beyond this, we have come to realize that regimes can have broader consequences, regardless of their effectiveness in dealing with the specific problems that motivate their creation (Underdal and Young 2004). In an anarchical social setting in which there are few explicit mechanisms to deal with social change at the systemic level in an orderly manner, in fact, the creation and operation of a growing collection of issue-specific regimes can play a significant—albeit de facto—role in addressing larger issues of social change. In short, there is no lack of issues on the research agenda of those interested in international regimes.

## 1.3 Regime Analysis: Challenges and Responses

The bulk of the research carried out to date on international regimes employs qualitative methods and, more often than not, directs attention to the development of case studies. Many analysts have sought to mine intensive case studies of individual regimes to develop larger conceptual and theoretical arguments about the formation and operation of institutions in international societies. We think immediately, in this connection, of the work of scholars like Peter Haas, who used a study of the environmental regime for the Mediterranean Sea in developing his idea of epistemic communities (Haas 1990); Karen Litfin, who explored the role of discourses and knowledge brokers through a study of the creation of the regime dealing with the stratospheric ozone layer (Litfin 1994); and Edward Parson, who focused on the ozone regime again, but this time as a means of presenting an argument about the role of scientific assessments (Parson 2003). Others have made use of a number of cases (usually three to fifteen) in an effort to develop and explore generic hypotheses about the formation and effectiveness of regimes. Robert Keohane and his colleagues drew on seven cases in developing the idea of the three Cs—increased concern, improved contractual environment, and enhanced capacity—as sources of regime effectiveness (Haas, Keohane, and Levy 1993). Oran Young and his team focused on three cases in seeking to differentiate among a number of behavioral mechanisms through which regimes influence behavior (Young 1999a). Edward Miles, Arild Underdal, and their colleagues were able to make use of fifteen cases in exploring a range of important issues like the degree to which the problems that regimes address can be classified as benign or malign in nature (Miles et al. 2002). But note that all these studies rely on discrete or self-contained case studies that are then used to draw comparisons, in contrast to the development of crosscutting hypotheses that are then "tested" against evidence drawn from a sizable universe of cases.

There is no question that such methods have served us well in the effort to improve our understanding of the nature and role of international regimes. Yet it is also apparent that these methods have more or less severe limitations as procedures for moving us into the next stage of regime analysis. Two such limitations seem particularly important. First, and perhaps most fundamentally, we lack standardization with regard to the definition of terms, the selection of variables, and the operationalization of hypotheses. This problem goes right to the heart of the enterprise in the sense that members of the research community do not agree fully regarding the meaning of the term *regime*. Whereas some analysts are prepared to say that a regime comes into existence with the signing of an agreement spelling out its terms, for example, others want to weed out arrangements that exist on paper but that do not give rise to observable social practices (Rittberger 1993). And this is only the beginning of the problem. While some scholars equate regime formation with the signing of a formal agreement setting forth the relevant rules and procedures, others are prepared to acknowledge the formation of regimes in the absence of such constitutive contracts. Similar problems arise when it comes to the selection of variables and the formulation of hypotheses. Whereas some analysts take a relatively narrow view of the idea of effectiveness and concentrate on matters like compliance, others think of effectiveness in terms of problem solving and express little interest in phenomena like compliance per se (Victor, Raustiala, and Skolnikoff 1998). For their part, hypotheses pointing to the roles that hegemonic actors, epistemic communities, and individual leaders play in processes of regime formation are hampered by disagreements about the operational meaning of key terms, a problem that makes it difficult or even impossible to frame these propositions in a manner that is falsifiable. Small wonder, then, that readers of the regime literature frequently come away with the feeling that the findings of individual studies are difficult to compare with one another, even though the terminology they employ seems compatible in superficial terms.

The other limitation arises in studies that seek to draw on a number of discrete cases to "test" hypotheses about the formation and operation of regimes. Commendable as it is to use a number of cases in this connection, most of these studies are constrained by the absence of a truly comparative mode of analysis. In the typical case, they involve a number of separate and self-contained case studies prepared by different individuals who share a commitment to the general goals of a project but who have not worked out clearcut definitions of key terms in advance and made a firm commitment to stick to a set of common terms and definitions throughout their work. The result is an effort to formulate a set of generalizations applicable across a number of cases that is plagued by the lack of a research design that guarantees the internal validity of such comparisons. No doubt, reaching agreement on such a research design and making a commitment to stick to it require a high level of discipline among the members of research teams. But it is disconcerting, to put it mildly, to examine the individual case studies that feed into projects of this kind and to realize that the persuasiveness of any general conclusions reached is jeopardized by problems of comparability across the case studies on which they rest.

What is to be done to mitigate these problems and, in the process, liberate regime theory to move into a new and more analytically rigorous mode? This is the point of departure for the contributions of the International Regimes Database (IRD), together with analyses based on the use of the database such as those reported in the substantive chapters of this book. The IRD is not a compilation of case studies in the ordinary sense of the term. Rather, it is based on a common data protocol that identifies and defines a large set of variables relevant to all members of the universe of international regimes. Transformed into an extensive questionnaire, the IRD data protocol has served as the instrument that expert coders use in providing data regarding many features or aspects of individual regimes. Although it would be possible to reconstruct a description of an individual regime from the information included in the database, the database itself takes the form of a set of tables containing information on all the cases included in the database, which users can query in their efforts to answer questions of a general nature regarding a wide range of issues. To the best of our knowledge, the IRD is the only tool of this sort currently available for systematic empirical research on international regimes. The results that we report in our substantive chapters therefore constitute an important step forward in efforts to evaluate hypotheses about international regimes.

## 1.4 Using the IRD: Epistemology, Theory, and Methodology

The IRD is an example of what is widely known among database managers as a relational database. Such a database contains a number of tables that can be compared with one another. The term *relational* as used here refers to the nature of the model that underlies the collection and manipulation of data in a database format. So far as we know, the choice of this database-management system does not introduce any biases likely to affect the results derived from using the IRD. Nevertheless, it is worth noting that the use of this model does have important implications for data structure, data integrity, and data manipulation (Whitehorn and Marklyn 2001, 211–213). Among the most significant of these is the fact that the IRD makes it possible to compare records on specific aspects of a number of regimes coded by contributors using a single, well-defined set of concepts, definitions, and scales.

We developed the IRD to reflect and to facilitate further analysis of the major theoretical issues dominating the study of regimes during the years leading up to the construction of the data protocol. This feature of the database has several important consequences. The protocol devotes a great deal of attention to regime attributes. This allows users both to focus on specific institutional attributes (e.g., decision rules) in order to investigate claims that regime design matters in the sense that institutional features affect the course of interactive decision making (Mitchell 1994), and to treat individual features as dependent variables in theories seeking to explain the substantive character of specific regimes (Koremenos, Lipson, and Snidal 2001). Beyond this, we sought to structure the database with the major theoretical debates in this field of study in mind. The section on regime formation, for instance, is organized in such a way as to facilitate studies of the roles that power, interests, and knowledge play in the processes through which regimes are created. Similarly, the section on regime consequences reflects a conscious effort to facilitate analyses of effectiveness that differentiate among outputs, outcomes, and impacts as distinct categories of effects arising from the operation of regimes.

It follows, of course, that the IRD could not and does not anticipate changes in the research agenda relating to international institutions during the period since its development. This is a significant limitation. For example, the IRD does not pay close attention to the phenomenon of interplay between or among distinct institutional arrangements, a topic that has become a growing area of analytic interest over the last few years (Young et al. 1999; Underdal and Young 2004; Oberthür and Gehring 2006). Similar comments are in order regarding some of the most recent thinking about the sources of compliance with international and transnational rules (Zürn and Joerges 2005). Yet it would be wrong to overemphasize this limitation, for several reasons. Studies of the formation of regimes and their effectiveness in dealing with the problems leading to their creation remain the core of the research agenda for those interested in the roles that regimes play as sources of governance at the international level. An emphasis on attributes, on this account, is also an element of the core, since many analysts are interested either in explaining how processes of regime formation lead to the choice of specific combinations of attributes or in investigating the links between the attributes of regimes and the consequences arising from their operation.

What is more, it is often feasible to make good use of the IRD in dealing with emerging research questions. Take the issue of institutional interplay as a case in point. Assuming that the analyst is able to devise measures of the results of such interactions in terms of criteria like synergy and conflict, it is easy to see how the IRD can be brought to bear in efforts to explore links between these results and the attributes of the regimes involved in institutional interplay. Does synergy increase, for example, when the regimes interacting with one another share certain basic design features? There are many cases, too, in which it will prove beneficial to link variables included in the IRD with any of a number of variables introduced from other domains. To illustrate, suppose an analyst wants to know whether regimes in which all the members are democracies or advanced industrial societies are more effective than regimes that are more heterogeneous in these terms. The proper strategy here is to combine outside indicators of democracy or industrial development with the IRD's measures of regime consequences. Because the possibilities for links of this sort are virtually unlimited, we made no effort to include data on factors of this nature in the IRD itself. But we have worked hard to make it easy to use data included in the IRD in exploring any number of inquiries of this type.

We should reemphasize as well the fact that the IRD is constructed in such a way as to accommodate data relating to regimes operating in a variety of issue areas. At this stage, the data included in the IRD all pertain to environmental arrangements. Yet the basic issues regarding matters like regime formation and effectiveness are generic; they are essentially the same whether we are examining regimes dealing with arms control, trade, and human rights or regimes governing human-environment interactions. It would be straightforward, under the circumstances, to augment the IRD with data relating to other types or categories of regimes. One attractive result of such an expansion would be to allow analysts to address questions of the following form: Do the links between the use of certain compliance mechanisms and the effectiveness of regimes differ as a function of the issue area in which the relevant regimes operate?

The IRD will prove most attractive to those desiring to go beyond self-contained case studies in order to "test" propositions regarding the formation and performance of regimes using relatively large numbers of records dealing with specific variables. As we have said, some of these propositions will be endogenous to the IRD in the sense that data on both dependent and independent variables are included in the database. Propositions linking any number of regime attributes to success measured in terms of goal attainment or problem change exemplify analyses of this type. In other cases, propositions will link one or more variables included in the database with any of a wide range of exogenous factors. Propositions regarding links between the economic or political attributes of the member states and the effectiveness of regimes are illustrative of this category.

In the substantive chapters of this book, we rely largely on descriptive statistics. That is, we ask questions like the following: (1) What proportion of the cases using specified compliance mechanisms feature high levels of compliance, and does the regime make a difference in producing such results? (2) What proportion of the cases rely on consensus rules, does problem solving occur in such cases, and do regimes make a difference in this connection? Without doubt, the IRD lends itself to analyses employing a range of more powerful types of statistical inference. We urge those interested in using the IRD as a research tool to move forward in applying more sophisticated statistical procedures in studies making use of this analytic tool.

The IRD will be of interest mainly to those whose work on international institutions is empirical in nature. The database is fully compatible with the observation that institutions are social artifacts rather than material entities. But it allows analysts to ask questions of the following sort. Does the allocation of issue-specific power among the participants affect the outcomes of institutional bargaining? Are regimes whose rules are legally binding on the members likely to prove more effective than those whose rules are more informal? Is there a difference between goal attainment and problem solving in terms of the determinants of regime effectiveness?

At the same time, we sought to construct the IRD in such a way that it would be of interest to analysts whose thinking is rooted in a variety of approaches to the study of international relations, including neorealism, neoliberal institutionalism, and constructivism. Neorealists can link data in the IRD pertaining to any of a number of features of regimes with exogenous measures of the distribution of power among the members of these arrangements. Neoliberal institutionalists can probe issues relating to the number of actors (or blocs) involved in regime formation or the roles that nonstate actors play in processes of regime formation. Constructivists are likely to find most interesting those data pertaining to knowledge about the nature of the problem, information available to those engaged in institutional bargaining, and awareness of the full range of possible responses to the problem. While the animating logic of the IRD is empiricist in nature, then, it should appeal to those whose thinking rests on a variety of analytic foundations and prove useful in probing some of the fundamental debates among adherents to a variety of theoretical perspectives on the nature of international relations.

Methodologically, the IRD's fundamental contribution is to promote and facilitate a transition from the case study to the relational database in research on international regimes. What does this mean? Because the IRD's basic unit of analysis is the regime element (see chapter 2 for a detailed explanation of the architecture of the database), it is easy to devise queries about specific variables included in the database that generate hundreds of records. If we launch a query regarding the links between the nature of decision rules and the attainment of stated goals, for instance, the database will produce a table that includes all records for all regime elements produced by all coders. Obviously, this does not completely eliminate concerns about internal validity. It is possible that the judgment of some coders is poor or that disagreements among coders will skew the results. Even so, this is a major step forward relative to earlier efforts to identify the determinants of regime formation and effectiveness relying on discrete case studies with little control over the use of common concepts, variables, and indicators (Young and Osherenko 1993; Haas, Keohane, and Levy 1993; Victor, Raustiala, and Skolnikoff 1998; Young 1999a; Miles et al. 2002).

At the same time, the IRD is not without interest for those whose work on international regimes is more qualitative in character. The database allows coders to record judgments about (un)certainty or complications that made it hard to respond to specific questions in the data protocol with respect to their case. In the section on compliance/conformance, for instance, the protocol asks coders to record the basis of their judgments (e.g., official documents, interviews, published materials) regarding each important actor. Similarly, the protocol allows coders to express their views regarding the causal mechanisms affecting the behavior of key actors with regard to compliance/conformance. The result is an opportunity to engage in qualitative analysis in a setting that makes it possible to use information relating to a range of cases and in which it is feasible to examine comparable information across individual cases. Although the most distinctive feature of the IRD is its ability to support the "testing" of hypotheses using large numbers of records, then, we believe the database will also appeal to analysts desiring to engage in qualitative studies that draw on information pertaining to a number of distinct regimes.

## 1.5 Plan of the Book

In the body of this book, we proceed as follows. Chapter 2 deals with a collection of more specific issues pertaining to the architecture of the International Regimes Database along with a variety of methodological concerns we encountered in the course of developing the database. Not only does it contain a more detailed account of the selection of variables, the negotiation of the precoding agreements, and interactions between the database manager and individual coders, it also explores a range of more technical issues dealing with coder reliability and data validity, the transformation of answers supplied by individual coders into electronically accessible data, and the choice of software used in the creation of the database. Readers who are mainly interested in considering the substantive results flowing from uses of the database may want to skim over some of these issues, especially on a first reading of the book. But those who are more technically inclined will find much of interest here; even those more interested in the substantive results are likely to find themselves turning back to the materials included in chapter 2 in order to answer any number of questions that enter their minds as they seek to understand and evaluate arguments set forth in the substantive chapters to follow.

Chapters 3–5 present a number of findings based on our initial efforts to make use of the IRD as an analytic tool. In each of these chapters, we draw on the IRD as a source of descriptive statistics that can tell us whether expectations derived from theoretical arguments are borne out in practice when we look at data drawn from a relatively large number of cases. Both because the issues are critical to the future of regime theory as a robust area of inquiry and because we have sought to produce a coherent and integrated volume, we have chosen to present preliminary findings that deal in one way or another with the consequences flowing from the operation of regimes.

Chapter 3 takes compliance, a kind of regime outcome, as the dependent variable. It proceeds to ask questions about the mechanisms that produce behavior that is (non)compliant with regime rules. How much of the variance in these terms is explainable in utilitarian terms or, in other words, in terms of benefit-cost calculations on the part of those subject to a regime's rules? What is the significance in this context of other behavioral mechanisms such as respect for legally binding obligations, the force of legitimacy, and the influence of what some analysts have described as managerial approaches to compliance (Chayes and Chayes 1995)? The goal here is not to attribute all the variance to a single mechanism. Rather, we want to understand how a number of mechanisms operating at the same time can account for observed results and, in the process, to illuminate a set of issues that have provoked a lively debate among students of international relations in recent years (Raustiala and Slaughter 2002).

Chapter 4 examines links between the existence and operation of both decision rules and compliance mechanisms and the observable consequences of regimes measured in terms of both goal fulfillment and problem solving. Are the results flowing from arrangements that allow some form of majority voting more likely to be effective in solving problems than the outcomes arising under consensus rules? What difference does the inclusion of mechanisms featuring notices of violations, suspension of membership rights, financial penalties, contributions to capacity building, and so forth make in terms of the actual consequences arising from the operation of a regime? This question ties into a larger theoretical debate among proponents of what have become known as collective-action and social-practice models of regimes (Young 2002c). Those who think in collective-action terms generally focus a lot of attention on matters of compliance and enforcement (Downs, Rocke, and Barsoom 1996). They expect to find important links between the operation of compliance mechanisms and actual consequences. Those more comfortable with social-practice models, by contrast, focus on the extent to which the behavior of regime members becomes routinized and subject to the influence of the logic of appropriateness (March and Olsen 1998). They expect to observe weaker links between the operation of compliance mechanisms and actual consequences. Of course, we cannot settle the differences between these influential schools of thought in a single chapter. But we can offer insights that bear directly on this issue.

Chapter 5 adopts a somewhat different perspective on the issue of regime consequences. Whereas chapter 4 focuses on regime attributes and their links to results, chapter 5 directs attention to a variety of programmatic activities ranging from monitoring to implementation review and assessments of the adequacy of existing commitments. This leads to an examination of a pair of major links in the processes through which regimes affect the outcomes of interactions among their members. The first link features the emergence of consensual knowledge out of programmatic activities; the second centers on the role of consensual knowledge as a determinant of the consequences arising from the operation of regimes. What is suggested here is that regimes often produce effects through a two-stage process running through the production of consensual knowledge and the application of that knowledge to influence the behavior of the actors involved in a given issue area. In this way, the influence of regimes goes deeper than simple accounts stressing the role of incentive mechanisms suggest.

Chapter 6 presents our own assessment of the IRD and of lessons learned in conjunction with this project that will be of interest to other theoretically inclined students of international relations. Of course, we are so close to the effort to construct the IRD that we may lack perspective on some of the advantages and disadvantages of the choices we have made. Outsiders, moreover, are apt to be better placed to adopt an unbiased or objective perspective on the results we have produced. Nevertheless, we have learned a lot in the long process of designing, developing, and using the IRD; we are still learning as we make the necessary arrangements to open the database for use on the part of others. There is nothing like the need to address issues in practical terms to sharpen understanding of the implications of a wide range of analytic and methodological matters. In developing the IRD, we had to make explicit decisions about a large number of matters; there was no option of glossing over such matters or putting off resolutions until some later time. Our purpose in this final chapter is to share with readers a sense of how we came to terms with these issues, our own judgments at this juncture regarding the value of the IRD, and some recommendations arising from this experience that may prove helpful to others embarking on large-scale and labor-intensive projects in the field of international relations.

Appendix A contains key excerpts from the IRD Data Protocol. This is a somewhat bulky document. But it will interest users of the IRD as well as those simply desiring to get a better sense of the range of theoretical concerns we sought to address in designing the database. This document, more than anything else we can offer, paints the big picture of the goals and objectives that guided us throughout the long process of making the transition from paper to practice in the construction of the IRD. The complete text of the IRD Data Protocol is included in the CD that comes with this volume. Appendix B describes a procedure for interrogating the database that constitutes an alternative to the construction of queries in

MS Access and allows the user to construct visual representations of the relationships among variables included in the IRD. In our view, the way forward lies in employing a number of analytic tools and comparing the results in search of patterns that emerge from work making use of several distinct approaches to the study of international regimes.

Finally, we have included with each copy of the book a CD containing all the data currently included in the database. Anyone who has a recent version of MS Access will be able to use these data to construct queries regarding any of the variables in the database. We hope that many researchers interested in international regimes will find this a useful tool in pursuing their studies. We will be happy to answer questions from those interested in using the IRD, and we look forward to seeing the results of their efforts. In this way, we are pleased to provide the IRD to members of the interested community as a pure public good.

# 2

# Database Architecture, Case Structures, and Coding Procedures

## 2.1 Introduction

A database is a means to an end, not an end in itself. The formulation of concepts and research questions should govern decisions regarding the inclusion of variables in a database. Our decisions regarding the architecture of the IRD emerged from a process involving the development and testing of several versions of the data protocol (Breitmeier et al. 1996a, 1996b). Along the way, we confronted a number of methodological issues that are well known to analysts who conduct qualitative case studies regarding the formation or effectiveness of governance systems in international society. We discussed many of these topics in depth during trial runs carried out with case-study experts to test our codebook and the conceptual framework to be used for the development of case designs. We gained important insights from these trial runs and developed pragmatic solutions to the methodological problems that emerged from this process. Numerous methodological issues arose as well in conjunction with the actual coding of cases for inclusion in the IRD.

Methodological discussions have a reputation for being boring; they frequently serve as ritual exercises that empirical analysts engage in to prove their methodological sophistication. Nevertheless, such a discussion is relevant here to provide information needed by members of the scientific community to understand or criticize our research design and to assess the reliability of our data (King, Keohane, and Verba 1994, 8). The discussion in this chapter also offers regime analysts practical solutions for a number of familiar conceptual and analytic problems.

Describing the guidelines we developed for the design of case structures may help others who face similar problems in developing case designs for empirical and especially comparative research on regimes.

In the body of this chapter, we focus on three methodological concerns that are central to our project. The first section deals with questions pertaining to the architecture of the IRD. How can we design a data protocol that reflects the major theoretical concerns and dominant research questions of regime analysis? Can we operationalize concepts to allow reliable measurements for key variables? Is it possible to include information regarding political dynamics affecting the different levels of interstate relations or the broader context of global society?

The next section addresses a number of conceptual issues that arose in developing the procedures needed to design or construct individual cases. What is a regime, and how can we separate individual regimes from other institutions? Can we devise a case design that takes the legal and institutional complexity of regimes into account? Is it feasible to consider factors governing the evolution of regimes over time? Can we code the actions of various types of actors active during the formation and evolution of regimes?

The final section considers issues relating to coding procedures and methods for analyzing the data included in the IRD. Are there procedures that can ensure the production of a reliable dataset? Are the questions, operationalizations, and explanations included in the data protocol sufficiently clear to allow experts from disciplines other than political science to code cases? What are the risks of various types of biases arising during the coding process? How can members of the research community make use of the IRD?

## 2.2 Database Architecture: From Concept to Design

The study of international regimes has developed into a substantial enterprise over the course of the last twenty-five years. In selecting variables for inclusion in the IRD, we considered the major theoretical issues and research questions that have motivated regime analysis during this period (Levy, Young, and Zürn 1995). Many variables were candidates for inclusion in the database. We concluded early on that it would be impossible—and ultimately unnecessary—to incorporate a wide range of socioeconomic and political factors of obvious importance to the creation and operation of regimes. It is easy to see that factors such as economic booms and busts, military actions, major demographic shifts, and the introduction of new technologies can have profound consequences not only for success in efforts to form regimes but also for the capacity of these arrangements to solve or manage specific problems. Nonetheless, any effort to include all factors of this sort that could affect the formation or the performance of regimes would lead immediately to an expansion of the database to unmanageable proportions; it might fail in any case to include some socioeconomic factor that turns out to be important in understanding a specific regime.

Fortunately, data covering most of these matters are relatively easy to obtain from independent sources. Accordingly, we quickly adopted the principle that users of the IRD will need to import data on a variety of socioeconomic and political factors needed to carry out specific studies of the creation, operation, or performance of international regimes. This principle assigns responsibility to individual users of the IRD for making choices about the relevance of factors relating to the broader social setting within which regimes operate and for mobilizing data needed to examine these factors. It constitutes a design feature essential to maintaining the coherence and usability of the database.

Regime analysis has been characterized by an expansion of research questions and the growth of efforts to consider different levels of analysis. Although many case studies deal with environmental regimes, empirical studies of regimes extend as well to all major issue areas of international affairs. For this reason, we have worked hard to create a data protocol usable to code regimes dealing not only with environmental matters but also a range of economic, security, human rights, and other prominent issues in world politics. While many forms in the IRD Data Protocol deal with variables and supplementary explanations needed to code environmental regimes, it would be easy to reorient the protocol to allow for the coding of regimes in other issue areas.

We have worked hard also to recognize the expansion of regime analysis from its original emphasis on regime formation to a growing interest in the effectiveness of regimes and patterns of change in regimes over time. In terms of architecture, therefore, we have built the IRD on four pillars, each of which includes "variables that address a major area of interest to students of international affairs": regime formation, regime attributes, regime consequences, and regime dynamics (Breitmeier et al. 1996a, 3).

Increasingly, analysts understand regimes as institutional arrangements that allow for participation by a broad range of actors. For this reason, we sought to develop a research tool capable of recognizing participation on the part of various types of nonstate actors as well as outcomes and impacts affecting transnational or subnational levels of social organization. During the design stage, we also sought to structure the data protocol to include variables that will allow users to explore and to test hypotheses about a wide range of theoretically significant perspectives on regimes. Thus, the data protocol asks for information about the roles of power, interests, and knowledge; the activities of states, nonstate actors, and individuals; the stages of agenda formation, negotiation, and operationalization; outputs, outcomes, and impacts; the effectiveness of regimes in solving specific problems, and the broader consequences of regimes.

The scope of the IRD is broad. Nonetheless, it is only fair to note that new conceptual and theoretical issues concerning international institutions have come into focus since we developed the architecture of the database. In the nature of things, we were unable to include new research interests emerging after the completion of the IRD Data Protocol. By addressing issues of formation, effectiveness, and change, we sought to cover the core concerns of regime analysis. We strove as well to be as forward-looking as possible in anticipating the emergence of new areas of interest among analysts studying international regimes.

What are the implications of this constraint? If we were designing the data protocol today, it would differ in some respects from the existing protocol. In chapter 6, we discuss several pragmatic ways to deal with this limitation. At this point, it will suffice to say that any changes we might introduce today would not alter the basic architecture of the database. Specifically, we would retain the four major sections on regime formation, regime attributes, regime consequences, and regime dynamics, adding new variables only within the basic framework of the database.

The finished protocol consists of 136 forms that call on coders to supply data about more than 200 variables. Not surprisingly, some coders—especially those assigned to regimes featuring several components or watersheds—found the task of responding to the queries included in the data protocol extremely challenging. It is not hard to think of questions pertaining to the formation, operation, and performance of international regimes that the IRD Data Protocol does not address; this is inevitable in a project of this sort. It is important to be candid about the limits of the database, acknowledging the existence of issues that are not included in the data protocol and recognizing the fact that newly emerging issues will not appear in the database. But it is equally important not to exaggerate this limitation. The IRD does cover the core issues that animate most studies of international regimes; the architecture of the IRD also creates opportunities for users to investigate some new issues by using variables that are included in the database as proxies for newly emerging issues.

### 2.2.1 Regime Formation

A number of efforts to identify the conditions governing regime formation from a comparative perspective preceded the database project (Young and Osherenko 1993; Rittberger 1990). These projects investigate "power," "interests," and "knowledge" as variables explaining the formation or evolution of regimes. The debate between neorealists and neoliberals about the relevance of international institutions in world politics has given further impetus to the construction of a database allowing researchers to check the validity of theoretical assumptions underlying these competing approaches (Keohane and Martin 1995; Mearsheimer 1994–1995; Strange 1983). Variables of interest to both camps are included in the data protocol.

Some may regard this section of the database as a bit outdated given the emergence of analytic questions dealing with issues like regime consequences and the design of regimes. Yet our understanding of the processes of regime formation and evolution over time remains limited. The factors that regime analysts emphasized in the 1980s to explain regime formation were based on a state-centric understanding of world politics. But the rise of transnational civil society suggests the need to explore whether state power has been constrained by transnational social forces and whether state interests are influenced by these transnational forces (Breitmeier and Rittberger 2000; Khagram, Riker, and Sikkink 2002; Take 2002; Beisheim 2004). This section of the protocol makes it possible to explore whether important states, nonstate actors, or individuals assumed roles as pushers or laggards during processes of regime formation; it also includes data relating to the strategies these actors used to achieve their political goals. The result is a dataset that describes the activities of a broad range of states, nonstate actors, and individuals in world politics.

Although we originally conceived of this section as a basket containing variables relevant for the study of regime formation per se, the trial runs revealed the need for information about dynamic processes affecting regimes following their initial creation. We have therefore framed the variables dealing with regime formation in such a way that analysts can also use them to examine political processes arising during the evolution of a regime. Specifically, we developed a case design that distinguishes temporal stages occurring after the initial creation and development of a regime.

The study of political processes involved in the creation and evolution of regimes is relevant to the explanation of certain outcomes and impacts of regimes. Thus, many of the power-, interest-, and knowledge-based approaches used to explain the formation or evolution of regimes apply in a slightly modified form as explanations for various aspects of regime effectiveness. Accordingly, the variables included in the section on regime formation have been used as well to characterize changes occurring once a regime has been established.

The complexity of theoretical perspectives confronted us with the task of disaggregating individual approaches into distinct variables in the IRD Data Protocol. How can we decompose interest- or knowledge-based approaches into sets of less complex variables suitable for measuring developments that occur in connection with the evolution of interests or cognitive processes? The decomposition of the knowledge-based approach, for example, was reflected in a number of variables in the IRD codebook dealing with the degree to which the nature of the problem was well understood; the availability of information about options for dealing with the problem; and the degree to which knowledge of the probable consequences of different options for solving the problem was well established.

Other variables make it possible to evaluate the relevance of actors that produce new scientific knowledge or disseminate knowledge to the public sphere. This section of the data protocol asks for information about the roles of intergovernmental organizations, epistemic communities, and individual states in the production and dissemination of consensual knowledge. It also seeks to measure the constellations of interests that emerge in specific issue areas. To this end, we developed procedures for identifying the relevance of conflict about the distribution of costs, determining the existence of incentives to disobey the rules after a regime is put in place, describing the relevance of complexity in the issue area, and assessing the degree of compatibility or incompatibility of interests among states in the issue area.

### 2.2.2 Regime Attributes

The second section of the data protocol includes an extensive set of variables dealing with regime attributes or, in other words, significant features of institutions that give individual regimes their distinct character. Many regime analysts have focused special attention on the role of institutional arrangements as mechanisms through which rational actors can avoid collective-action problems (Keohane 1984; Haas, Keohane, and Levy 1993). Others have chosen a more complex approach to exploring regime effectiveness, considering institutions not only as mechanisms through which unitary actors behave as utility maximizers but also as arrangements involving different behavioral pathways, including norm-driven behavior, social learning, and the restructuring of social roles to improve outcomes and impacts (Young 1999a). This section of the data protocol offers a broad range of variables usable in efforts to test explanations that emphasize the different mechanisms through which regimes influence outcomes.

This strand of research rests on the assumption that by improving consensual knowledge or developing incentives to comply with international rules, institutions can prevent the overexploitation of natural resources or avoid the degradation of the environment caused by pollution (Chayes and Chayes 1995; Koremenos, Lipson, and Snidal 2001; Mitchell 1994a; Ostrom 1990; Weiss and Jacobson 1998; Victor, Raustiala, and Skolnikoff 1998). We can use information about regime attributes to "test" hypotheses explaining improvements in the knowledge base, variations in compliance on the part of members, or the effectiveness of regimes in solving the problems that give rise to their creation. The resulting knowledge will be relevant to broader political debates that focus on identifying institutional design features necessary for the management of transboundary problems (Young 2002c).

The regime-attribute section of the data protocol collects information about procedures and practices used to explore the nature of problems, monitor behavior, verify compliance, review implementation, and arrange financial and technology transfers. It also allows the analyst to identify the existence and operation of important regime bodies or the operation of procedures used to make collective choices.

The IRD's emphasis on institutional attributes raises a question about whether our project has an institutional bias, even though we claim that preferential treatment should not be given to any of the theoretical approaches that students of global governance use. We do not find this charge persuasive. Neorealists do not neglect the existence of institutions; they treat outcomes and impacts as predominantly a matter of translating structural power into bargaining leverage. Accordingly, they wish to understand the functioning of institutions to explain the role of structural power as a critical variable. A finding that structural power dominates political processes would certainly undermine constructivist claims that international rules and norms produce settings in which states are socialized in ways that serve to constrain the autonomy of individual actors in international politics. The IRD Data Protocol is a tool that analysts can use to assess the claims of any of these lines of analysis. It is clear as well that proof of the possible (in)significance of an institution can only be produced if major institutional attributes are taken into account during the coding process.

Our treatment of approaches used to explain compliance with regime rules illustrates that neoliberal and neorealist assumptions are treated in a balanced fashion. The enforcement approach to compliance gets equal billing in our codebook with the management approach. Emphasizing enforcement reflects a neorealist conception of international politics, whereas the idea of management has been developed by observers who argue that coercive pressure or sanctions are frequently ineffective and even irrelevant for achieving compliance, a conclusion suggesting that institutions should provide mechanisms that focus on capacity building in states lacking the resources needed to implement international rules and norms domestically (Downs, Rocke, and Barsoom 1996; Chayes and Chayes 1995).

This section pays special attention to identifying the substantive character of a regime's principles, norms, and rules as well as important goals agreed on among the members of a regime. Identifying these features of a regime requires expert coders to study legal documents (e.g., conventions, protocols, annexes, amendments, or decisions by COPs) in which these goals, principles, norms, rules, and decision-making procedures are established or where regime members have modified existing arrangements or added new institutional mechanisms. The database asks for information about these legal and organizational attributes for different time periods in the life of a single regime, so that the analyst can explore whether an expansion of institutional mechanisms or an increase in the scope, density, or specificity of regime rules has occurred over time. Such considerations are relevant to the broader debate about institutional change and the impact of such changes on efforts to solve the problems that lead to the creation of regimes in the first place (North 1990).

Although the preceding discussion directs attention to the roles institutional attributes play as determinants of regime effectiveness, analysts can also use these data to explore the conditions under which certain types of institutions emerge. Thus, they can examine links between the activities of various types of actors covered in the section on regime formation and certain regime attributes. To illustrate, we can ask whether the rise of nonstate actors as a relevant political force in world politics is changing the character of the principles, norms, rules, or decision-making procedures embedded in a variety of institutional arrangements.

### 2.2.3 Regime Consequences

Research on the effectiveness of international regimes has focused on efforts to measure the consequences arising from the functioning of regimes. Although this section of the data protocol includes the major consequences of interest to lawyers and political scientists, it is weaker with regard to economic concerns involving issues like the efficiency or cost-effectiveness of regime activities. This reflects the fact that reliable data needed to make calculations of costs and benefits are generally unavailable. Similar difficulties affecting this section arose with regard to a number of other regime consequences that case-study analysts have not examined systematically. At the same time, the data protocol does ask for information on a number of broader regime consequences (e.g., the impacts of regimes on neighboring issue areas, on overall relations among participating states, and on the character of international society as a whole) in the form of narratives usable in qualitative analyses.

The IRD includes ordinal scales regarding the extent to which the state of the world worsened, stayed the same, or improved as a result of regime activities with respect to several distinct topics of interest to students of regime effectiveness. We can assess the legal effectiveness of a regime with reference to variables in the data protocol that ask whether regime bodies are in operation and generate authoritative decisions, as well as whether important members have taken the steps needed to translate international commitments into domestic obligations. Measurements related to these legal matters can clarify whether programmatic activities are occurring as planned and decision-making procedures are functioning properly.

Compliance treated as a measure of behavioral effectiveness has gained prominence in recent research on international regimes (Mitchell 1994a; Underdal 1998; Weiss and Jacobson 1998). The data protocol contains two different approaches to compliance: (1) an overall judgment regarding conformance with regime requirements and prohibitions on the part of all subjects, and (2) a measure of individual conformance with norms and rules on the part of a number of the most important subjects. This distinction allows us to estimate the general level of compliance with a regime component as well as to create a dataset dealing with the compliance of single states.

A third measure of effectiveness focuses on changes in the cognitive setting (P. Haas 1990, 1992). Such changes may involve improvements in knowledge of the nature of the problem or in information about the options available for addressing the problem. The data protocol measures these factors with regard to important states. The resultant dataset is useful for exploring the impact of a regime on the attitudes of policymakers and the extent to which cognitive changes in important countries affect their behavior as pushers or laggards in regime formation or the levels of compliance they exhibit once regimes are in place.

Finally, and perhaps most importantly, two variables in the data protocol ask for information on whether goals agreed on in legal agreements have been attained and on the level of problem solving occurring under the terms of a regime component. These measures focus on goals and problems that expert coders identify prior to the start of coding. Goal attainment and problem solving are distinct phenomena. Goal attainment refers to progress in meeting goals articulated in formal agreements. While this variable reflects the extent to which the goals that policymakers agree on are fulfilled, it tells us little about the degree to which problems are solved (Bernauer 1995). The concept of problem solving reflects a more demanding standard; it refers to changes in the state of the problems identified during processes of regime formation or, in other words, the extent to which real-world conditions have improved during a specific time period. In using problem solving as a variable, we had to make a decision whether to treat the *collective optimum* as a standard against which to assess observed results or to focus on *relative improvements*. The idea of the collective optimum is more ambitious; it provides an ideal standard against which to measure progress (Helm and Sprinz 1999). But our trial runs revealed that it is often impossible for expert coders to identify the collective optimum in specific cases. As a result, we chose the less ambitious criterion as a baseline for assessments of the contributions regimes make to problem solving.

A central question in research on regime effectiveness concerns the degree to which international institutions actually cause observed impacts, or whether external forces like exercises of power or the introduction of new technologies are the real driving forces in world affairs. In many cases, observed results reflect the combined effects of regimes and external factors operating at the same time. In developing the section of the data protocol on effectiveness, we realized that several steps are involved in making reliable judgments about the impacts of regimes on observed levels of compliance, changes in the cognitive setting, goal attainment, and problem solving.

In the end, we chose to separate assessments of compliance, goal attainment, and problem solving from judgments regarding the causal significance of regimes. The latter require not only an assessment of the effects of nonregime factors but also a consideration of counterfactuals or the likely course of events that would have occurred in the absence of a regime (Fearon 1996; Tetlock and Belkin 1996). Counterfactual reasoning poses severe methodological problems. Modeling alternative pathways of social behavior can lead to arbitrary results when relatively long time periods are involved or when the analyst must control for a complex set of variables. Causal judgments alone cannot serve as measures of the outcomes and impacts of regimes. We have sought to supplement such judgments by identifying correlations between variables pertaining to regime attributes and changes in compliance, cognitive settings, goal attainment, and problem solving in order to generate additional evidence about the causal impacts of regimes.

### 2.2.4 Regime Dynamics

Regime analysis is underdeveloped with respect to the treatment of institutional dynamics. Neorealists and neoliberals express conflicting views, for instance, about the vulnerability of regimes to external disturbances, whether the resultant pressures involve developments in the issue area concerned (e.g., rising concentrations of greenhouse gases in the earth's atmosphere) or outside the issue area in such forms as business cycles and shifts in political power (Hasenclever, Mayer, and Rittberger 1997). The final section of the data protocol includes information about forces leading to changes in legal frameworks, the operation of institutional mechanisms, and the behavior of both members and nonmembers in the relevant issue area.

The limited scope of this section reflects the fact that we can derive information about many dynamic processes by examining relevant variables in the other sections of the data protocol. For example, because we have divided the evolution of regimes over time into several stages, examining changes occurring from one stage to the next reveals a number of dynamic processes, such as the extent to which new rules are added in the course of time or decision-making procedures are adjusted to deal with new issues. Such comparisons can also serve as a starting point for explorations of the expansion of programs for scientific research, the role of systems of implementation review, or the links between the development of effective methods of verification and the outcomes and impacts included in the section of the data protocol on regime consequences.

The idea of regime dynamics can refer as well to changing relationships between states and nonstate actors in political processes like agenda setting and institutional bargaining. A comparison of developments occurring during different stages of the lives of regimes can shed light on claims concerning the rise of global civil society and whether this development has influenced the effectiveness of individual regimes. A study of changes over time may also help in evaluating arguments about the relationship between institutionalization and regime effectiveness. Regimes that seem moribund in their early stages sometimes become more effective with the passage of time. Conversely, arrangements that seem dynamic in terms of the evolution of norms, rules, and decision-making procedures may fail to produce clearcut solutions to problems over time.

## 2.3 Structuring Cases for Regime Analysis

The international regimes included in the IRD rest on one or more legal documents (e.g., treaties, conventions, ministerial declarations). Our decision to focus on these so-called hard-law regimes reflects the fact that scientific debate has focused primarily on the role of institutions of this type. Analysts have conceived of these regimes as central elements of an approach to global environmental governance in which states make commitments to respect rights and to comply with norms and rules they have accepted as legally binding by taking steps to give them legal status within their domestic jurisdictions. While all twenty-three of the regimes currently included in the IRD feature one or more legally binding agreements, however, many of these arrangements encompass a mix of hard-law and soft-law elements. The 1987 Rhine Action Plan adopted by the riparian states in the aftermath of the 1986 Sandoz accident, for example, is a soft-law agreement in which states set ambitious targets for the reduction of pollution associated with dozens of substances. But the Conventions against Chemical Pollution and against Chloride Pollution, both adopted in 1976, established an institutional setting for the implementation of the action plan (Bernauer 1996). Analytic distinctions between hard- and soft-law regimes are often too crude to be helpful; they disregard the frequent interplay between the two types of arrangement in one and the same institution.

The complexity facing the analyst endeavoring to structure regimes for coding is caused not only by the variety of legal agreements associated with an institution but also by the fact that international governance systems encompass complexes of social relationships and political interactions occurring among regime members or arising from participation on the part of a range of nonstate actors in transnational political processes. As a result, the IRD team confronted the question of whether coding should address the regime as a whole or seek to reduce complexity by focusing attention on individual elements of regimes. The institutional patchwork of the GATT/WTO is so complex, for instance, that it is helpful for purposes of understanding patterns of governance to move beyond the macrolevel and consider separately the various components of the regime governing world trade.

In the beginning, we assumed casually that the individual regime would be our basic unit of analysis. But experience gained through the trial runs made it clear to us that this assumption was problematic. What was needed, we came to realize, was a set of criteria dealing not only with a regime's external boundaries but also with internal boundaries among a regime's components. Although it is obviously important to clearly identify periods of regime formation and termination, we found it relatively easy to identify these external boundaries. The specification of internal boundaries is equally important in analytic terms; it allows coders to explore important differences among the components of a single regime. But the decision to disaggregate regimes into their component parts made it essential to reach agreement with coders regarding the structure of a regime before actual coding began.

We used the resultant precoding negotiations to reach agreement on a common structure for each regime included in the IRD. Given our rule that two experts should code each regime and produce independent datasets for each case, precoding negotiations created shared understandings among the coders and the members of the research team regarding matters of case structure. These negotiations resulted in what we have come to call a precoding agreement setting forth common understandings concerning the nature of the problem(s) a regime addresses, the external and internal boundaries of the regime, and the identity of important actors in the regime (see appendix A). These negotiations were simple with respect to some regimes but long and arduous with respect to others. In every case, however, the precoding agreement provides some assurance that individual regimes are understood in the same way by the coders themselves and the members of the IRD team. This has proven critical in providing the IRD with a stable base.

The precoding process encompasses three major steps. Because international regimes are problem driven in the sense that their members normally create them to address problems they cannot resolve on their own, we sought at the outset to reach agreement on the identity and nature of the relevant problem. This allows coders to respond appropriately to questions in the section of the data protocol on regime consequences that deal with goal attainment and problem solving. Next, we negotiated agreement regarding a regime's internal boundaries. In this connection, we endeavored to reach consensus both about the identity of a regime's differentiable components and about what we now call temporal watersheds that separate different periods or stages in a regime's life. Combining distinctions pertaining to regime components and temporal watersheds allows us to identify the "regime element," the smallest unit of analysis in the IRD. Finally, we sought agreement about the identity of the most important actors for each regime and about the information regarding these actors to be included in the database. Of course, it is conceivable that biases will affect the outcomes of the precoding negotiations. But the fact that three or four experts have reached agreement regarding the terms of the case structure for each regime included in the IRD gives us a good deal of confidence in the validity of the results.

### 2.3.1 Problem Definition and the Specification of External Boundaries

A major feature of the precoding process for each regime was the development of an agreement regarding the nature of the problem(s) addressed. This ensured that the coders were able to supply information concerning the "same" regime. In the case of the regime created under the terms of the 1973 London Convention, for instance, the coders agreed that the purpose of the regime is to "prevent pollution by dumping of waste and other matter creating hazards to human health, harming living resources/marine life, damaging amenities or interfering with other legitimate uses of the sea."

Although this example seems straightforward, reaching agreement on the definition of a problem can be a tricky process. Problems are socially constructed; they involve human actions affecting the environment or conflicts that determine the situation structure of an issue area (Zürn 1992). The concept of situation structure directs attention to differences among key actors regarding both their role in causing a problem and their vulnerability to the impacts of the problem. In the case of the regime for the protection of the Rhine River, for example, coders agreed to define the problem as "pollution causing damage to ecosystems and water quality especially in downstream countries." Coders adopted a similar perspective on the regime for the Danube River, where the problem is described as "prevention/control of pollution from, in particular, hazardous substances and nutrients by up- and downstream countries into the aquatic environment of the Danube River Basin and into the Black Sea." Table 2.1 summarizes the agreed-on definitions of the problems addressed by all the regimes currently included in the IRD.

The problems environmental and resource regimes address often involve tensions among individual members regarding the relative weights to assign to distinct concerns (e.g., harvesting versus conservation). As a result, coders sometimes found it necessary to refer to several distinct, albeit related, concerns in the course of reaching agreement on problem definition. The cases of CITES and the regime for tropical timber exemplify this situation. Here, economic concerns, such as the "maintenance of a sustainable, legal trade in plants and animals" or "development of a commercially viable tropical timber industry," vie with concerns for conservation, such as the "protection of endangered species" or "minimizing

**Table 2.1**
Problems managed by regimes included in the IRD

| Regime | Problem |
|---|---|
| Antarctic regime | Growth of interest in managing exploitation of resources in and around Antarctica<br>Jurisdictional differences/conflicts about overlapping claims on the part of claimant states<br>Jurisdictional differences/conflicts between claimants and major nonclaimant states |
| Baltic Sea Regime | Concern for the state of the Baltic Sea as a very vulnerable and sensitive area |
| Barents Sea Fisheries Regime | Overfishing caused by strong competition over scarce fish stocks shared in the Barents Sea region |
| Biodiversity Regime | Conservation of biodiversity, sustainable use of its components, and fair sharing of benefits arising out of the use of genetic resources |
| CITES Regime (Trade in Endangered Species) | Maintaining a sustainable, legal trade in plants and animals<br>Protecting endangered species |
| Climate Change Regime | Increase of greenhouse gases in the atmosphere and inability of humans and ecosystems to adapt to the impacts of climate changes that ensue |
| Danube River Protection Regime | Prevention/control of pollution from, in particular, hazardous substances and nutrients by up- and downstream countries into the aquatic environment of the Danube River Basin and into the Black Sea |
| Desertification Regime | Land degradation in arid, semiarid, and dry subhumid areas resulting from various factors, including climatic variations and human activities |
| Great Lakes Management Regime | Managing a variety of human activities affecting the Great Lakes themselves and the regional ecosystem of which the lakes are the core |
| Hazardous Waste Regime | Exports and imports of hazardous waste from industrialized to developing countries |

**Table 2.1**
(continued)

| Regime | Problem |
|---|---|
| IATTC-Regime (Inter-American Tropical Tuna Convention) | To conduct scientific studies on, and make recommendations for, the conservation of dolphins captured by vessels fishing tuna in the eastern Pacific Ocean<br>To conduct scientific studies on, and make management and conservation recommendations for, tuna, baitfishes, and other kinds of fish taken by tuna fishing vessels in the eastern Pacific Ocean |
| ICCAT-Regime (Conservation of Atlantic Tunas) | To ensure effective international conservation and management of tunas and tunalike species that migrate extensively in the Atlantic Ocean, including the adjacent seas |
| Regime for the International Regulation of Whaling | Orderly development of the whaling industry<br>Conservation of whale stocks |
| London Convention Regime | To prevent pollution by dumping of waste and other matter creating hazards to human health, harming living resources/marine life, damaging amenities, or interfering with other legitimate uses of seas |
| ECE-Regime on Long-Range Transboundary Air Pollution | Transboundary air pollution causing damage to ecosystems<br>Enhancement of East-West cooperation and détente |
| North Sea Regime | The protection of the marine environment of the northeast Atlantic, with emphasis on the North Sea area |
| Oil Pollution Regime | Coastal oil pollution and seabird deaths due to oil pollution from intentional discharges and accidental oil pollution |
| Regime for Protection of the Rhine against Pollution | Pollution causing damage to ecosystems and water quality, especially in downstream countries |
| Ramsar Regime on Wetlands | Wetlands as components in human development, conservation of biodiversity and water issues<br>Wetland reclamation, destruction and degradation, and the impact of this habitat loss on the conservation status of wild birds |

**Table 2.1**
(continued)

| Regime | Problem |
|---|---|
| Regime for Protection of the Black Sea | Ongoing degradation to the ecosystem of the Black Sea and unsustainable use of its natural resources |
| South Pacific Fisheries Forum Agency Regime | Coordination of fisheries management among the members of the SPF in order to regulate tuna harvest by distant water fishing nations, and to maximize returns to the Pacific Island countries |
| Stratospheric Ozone Regime | Losses of stratospheric ozone caused by ozone-depleting substances |
| Tropical Timber Trade Regime | Underdevelopment of a commercially viable tropical timber industry<br>Increased evidence of significant levels of tropical deforestation and rainforest degradation in developing tropically forested countries |

tropical deforestation and rainforest degradation in tropically forested developing countries" (Sand 1997; Gale 1998).

In other regimes, coding experts decided to include competing economic and environmental interests as elements of a single problem definition. A case in point is the biodiversity regime where the coders agreed on "the conservation of biodiversity, sustainable use of its components, and fair sharing of benefits arising out of the use of genetic resources" as a characterization of the problem addressed by this regime. While the coders agreed on a single problem for the majority of regimes, there are cases involving several distinct problems. The Antarctic Treaty System, for example, addresses the "growth of interest in managing exploitation of resources in and around Antarctica." But this regime also deals with jurisdictional matters arising from conflicts among claimant states and between claimants and major nonclaimant states (Joyner 1998).

The perceptions of regime members concerning the problems at stake can change and may diverge over time. Some changes of this sort are so dramatic that they result in the definition of new problems. Consider developments in the IATTC regime as a case in point. In its initial form, this regime focused on the conduct of studies and the development of recommendations for the conservation of tunas, baitfish, and other species caught by tuna boats in the Eastern Pacific Ocean. Although the regime still deals with these issues, concern has grown since the mid-1970s about the killing of dolphins in conjunction with the harvesting of tuna in this region. On the other hand, the regime dealing with long-range transboundary air pollution in Europe is a case in which a problem regarded as important at the outset has faded into the background. The damage to ecosystems caused by transboundary air pollution has remained a major concern throughout the life of this regime. But the enhancement of East-West cooperation, regarded as a major issue during the regime's initial phase, declined in importance over time (Levy 1993; Wettestad 2002).

Linkages among ecosystems can lead to problem definitions of an artificial nature. Although there are separate regimes for ozone depletion and climate change, for example, the relevant problems are connected (Oberthür 2001). Phasing out the production and consumption of ozone-depleting substances leads as well to reductions in emissions of greenhouse gases. Reductions in river pollution, to take another example, can help to solve problems centering on the pollution of regional seas. Thus, regimes for river basins like the Rhine or the Danube play a role in reducing land-based pollution of the North Sea and the Black Sea. We cannot ignore such interconnections among regimes. Still, it is apparent that we must establish boundaries between issues and the regimes associated with them in order to develop tractable units of analysis.

Problems often arise long before states reach agreement on the creation of regimes to cope with them. An important feature of the precoding negotiations, therefore, was the establishment of a starting date for each regime. Ultimately, we dated each regime to the signing of an agreement—legally binding or not—setting forth the constitutive provisions of the relevant arrangement. Often, constitutive agreements take the form of framework conventions or basic treaties governing interactions among those involved in the relevant issue area. Agreements like the 1979 Convention on Long-Range Transboundary Air Pollution, the 1985 Vienna Convention for the Protection of the Ozone Layer, and the

1992 Framework Convention on Climate Change mark the inception of regimes that have evolved subsequently through the addition of protocols, amendments, and more informal adjustments in practices (Levy 1995; Parson and Greene 1995; Oberthür and Ott 1999; Parson 2003). Although these conventions lack far-reaching commitments regarding the reduction of pollutants, they create a framework for future cooperation in the relevant issue area.

The development of the regime for the protection of the Danube River provides a case in which a soft-law agreement marks the starting point of a regime. Thus, the 1985 Bucharest Declaration of the Danube Countries to Cooperate Concerning the Water Management of the Danube triggered the establishment of several organizations designed to coordinate efforts to protect this river basin. Under the circumstances, the 1994 Convention on Cooperation for the Protection and Sustainable Use of the Danube River Basin transformed the regime into a legally binding arrangement and strengthened cooperation among the member states (Kaspar 1999).

Some regimes have more than one plausible starting point. And regimes can go out of existence before a new regime dealing with the same problem arises. The whaling regime created under the terms of the 1946 International Convention for the Regulation of Whaling was preceded by a convention negotiated in the 1930s whose provisions were weak and generally ignored. In this connection, we concluded that the termination of a regime occurs when (1) there is a temporal discontinuity in the operation of the regime, and (2) there is no identifiable successor or the eventual successor differs in fundamental ways from the original regime. Using this formula, we can date the existing whaling regime to the adoption of the 1946 convention.

The year 1998 marks the end point of the data coded for the IRD. Partly, the selection of this terminal date is attributable to the fact that the coding of individual cases started at that time. In part, the choice of 1998 stems from the desirability of choosing a common end point for regimes included in the database. In effect, this provides us with a basis for comparative analysis that is not distorted by artificial asymmetries in the datasets associated with individual regimes.

### 2.3.2 Identifying Internal Boundaries

In the course of the trial runs, we discovered that coders often have substantially different conceptions of the scope or content of individual regimes. In the end, we concluded that two phenomena account for most of these differences: the existence of distinct regime components and the occurrence of watershed changes or major transitions in the life of regimes. While some regimes (e.g., the Barents Sea fisheries regime) do not have distinct components, others (e.g., the long-range transboundary air pollution regime in Europe, the Antarctic Treaty System) encompass separate components dealing either with different sources of the problem or with different sets of regime activities.

We developed the data protocol in several stages aided by a number of experts who participated in trial runs that led to revisions of earlier versions. In the final version, components are associated with the existence of distinct institutional forms (e.g., separate treaties, protocols, annexes). But we supplemented this distinction with several additional criteria to avoid a situation in which each new legally binding agreement automatically triggers the development of a new component. In practical terms, the use of these criteria prevented the data protocol from becoming intractable for coders. It has also played a role in distinguishing important developments from more minor adjustments.

Three criteria for identifying separate components became especially relevant during the precoding negotiations. First, some components involve distinct institutional forms covering differentiable sources of the problem(s). The long-range transboundary air pollution regime, for example, has separate protocols dealing with different pollutants (e.g., sulfur dioxide, nitrogen oxides, volatile organic compounds, persistent organic pollutants) that have been added to the 1979 framework convention at various intervals. Second, we identified components where elements of regimes are aimed at different sets of activities or regulatory targets. The global trade regime, for instance, features different rules for trade in manufactured goods, agricultural goods, services, and so forth. Similar distinctions occur in environmental regimes. Thus, the basic Antarctic Treaty is supplemented by conventions dealing with the conservation of seals and the protection of Antarctic marine living resources as well as by the Protocol on Environmental Protection (Stokke and

Vidas 1996). Third, we identified components where distinct institutional forms address different regime functions (e.g., compliance, monitoring, funding). Although every regime includes provisions dealing with different functions, some regimes (e.g., those pertaining to ozone and climate change) have special institutional arrangements dealing with funding and technology transfers. Thus, we have treated the Montreal Protocol Multilateral Fund and the funding arrangement associated with the climate regime as distinct components.

We initially thought of two other factors as relevant to the identification of regime components: (1) arrangements that target different clusters of actors (e.g., developed versus developing states), and (2) provisions dealing with distinct subproblems (e.g., emissions of greenhouse gases from transportation systems versus industrial plants). But the expert coders did not find these distinctions helpful during the precoding negotiations.

Beyond this, we learned during the course of the trial runs that temporal distinctions are important. To be specific, it became apparent that individual regime components often experience major changes over time. Accordingly, we looked for "watershed" events that divide regimes into fundamentally different temporal stages and that therefore make it possible to subdivide regime components into smaller units of analysis.

Three types of watersheds emerged during the course of the precoding negotiations. A watershed occurs when there is a major restructuring of key principles and norms. In the case of Antarctica, for instance, the adoption of the Protocol on Environmental Protection marked a transition from resource management to environmental protection as the regime's fundamental mission. We also identified watersheds in cases involving changes in the group of leading actors. The ozone regime, for example, experienced such a change in the early 1990s, when the creation of a funding mechanism persuaded China and India to join the regime. In addition, watersheds occur when there are significant expansions in the functional scope of regimes (e.g., a radical deepening of regulatory rules). Such developments are relatively common in arrangements that evolve through the addition of substantive protocols to a framework convention.

Only seven of the twenty-three regimes included in the IRD experienced no temporal watershed. Many watersheds occurred during the 1980s and 1990s, a fact that reflects the dynamism of environmental policymaking during those decades. Such concerns found expression not only in broad perspectives on environmental problems (e.g., the overview articulated in the 1987 report of the Brundtland Commission and the overarching blueprint contained in *Agenda 21* adopted at the UN Conference on Environment and Development in 1992) but also in the creation and development of specific regimes.

Following an extensive debate, we decided to ask coders to provide separate answers to the questions included in the data protocol for each regime component. A similar story unfolded with regard to the occurrence of watersheds. Here, too, we asked coders to respond to the data protocol's questions separately for the periods before and after the occurrence of a watershed. While some regimes are simple arrangements featuring a single component that remains unchanged over time, others are far more complex. There are regimes in the IRD requiring coders to provide eight to ten separate answers to the same question. An obvious consequence of these decisions is a substantial increase in the number of "cases" included in the IRD. At this writing, the database contains data pertaining to 23 different regimes giving rise to a larger number of cases or "regime elements," when all components and watersheds are taken into account. A set of data comprising 184 regime elements would have emerged if all regimes were coded by two experts independently of one another. Because double coding exists for only 21 of the 23 regimes we have been able to include, the IRD contains data on 172 regime elements. Table 2.2 summarizes the case structures of the 23 regimes currently included in the IRD. It contains information about the starting date of each regime, the separation of regimes into components, and the occurrence of watersheds.

### 2.3.3 Roles of States, Nonstate Actors, and Individuals

Many parts of the data protocol adopt an actor-oriented perspective. This is especially true for the sections dealing with the formation or evolution of regimes and with regime consequences. Accordingly, those interested in the role of power can use the IRD to explore whether the actions of an important state or group of states caused others active in the issue area to agree to the formation of a regime or to its evolution over time. For their part, constructivists can use the database to ask whether changes in the cognitive setting affect the behavior of important states and whether states or other actors have tried to advance new modes of thought or worldviews to influence the behavior of other actors in an issue area. Those who employ a rationalist approach in analyzing regimes may be interested in exploring whether the formation and evolution of regimes occur primarily as a result of interstate bargaining and whether the operation of regimes helps to reduce uncertainties that states face in connection with issues of global governance.

The IRD contains information regarding the behavior of more than fifty states as well as the European Union. But states are not represented equally in this dataset. More than half of the entries relating to important states involve industrialized countries or countries with economies in transition. Some, like the United States, Russia (including the former Soviet Union), and Germany, are important actors in many regimes. The coders identified the United States, for instance, as an important actor in fifteen regimes, a fact that gives rise to a substantial dataset about American political behavior, compliance, and responses to changes in cognitive settings. Germany appears as an important actor in nine cases; both the European Union and Russia/USSR are included in six cases. Many developing countries appear as important actors in only one regime, although coders included Brazil in five regimes and China and India in three.

Coders treated only a few nonstate actors as important in more than one regime. IUCN, WWF, Greenpeace, and Friends of the Earth show up in several regimes, as do intergovernmental organizations like UNEP and the World Bank. These actors engaged in a wide range of activities relevant to the implementation of the provisions of environmental governance systems as well as to regime formation. But the great majority of the more than 100 nonstate actors emerged as important players in only a single regime.

Much the same is true regarding individuals identified as important actors during the precoding negotiations. The influence of most individuals is associated with the roles they play in specific regimes; only a

**Table 2.2**
The evaluation of case structures for twenty-three environmental regimes

| Regime | Regime component | Formation (Year) | Watershed (Year) | Second watershed (Year) | Endpoint of coding (Year) | Regime element |
|---|---|---|---|---|---|---|
| Antarctic Regime 1959–1998 | Antarctic Treaty | 1959 | 1980 | | | Antarctic Treaty 1959–1980 |
| | | | 1980 | 1989/1991 | | Antarctic Treaty 1980s |
| | | | | 1989/1991 | 1998 | Antarctic Treaty 1989/1991–1998 |
| | Convention of the Conservation of Antarctic Maritime Living Resources (CCAMLR) | 1980 | | 1989/1991 | | CCAMLR 1980s |
| | | | | 1989/1991 | 1998 | CCAMLR 1989/1991–1998 |
| | Convention on the Conservation of Flora and Fauna | 1964 | 1980 | | | Conservation of Flora and Fauna 1964–1980 |
| | | | 1980 | 1989/1991 | | Conservation of Flora and Fauna 1980s |
| | | | | 1989/1991 | 1998 | Conservation of Flora and Fauna 1989/1991–1998 |

| | | | | | | |
|---|---|---|---|---|---|---|
| | Convention on the Conservation of Seals | 1972 | 1980 | | | Conservation of Seals 1972–1980 |
| | | | 1980 | 1989/1991 | | Conservation of Seals 1980s |
| | | | | 1989/1991 | 1989 | Conservation of Seals 1989/1991–1998 |
| | Protocol on Environmental Protection | 1991 | | | 1998 | Protocol on Environmental Protection 1991–1998 |
| Baltic Sea Regime 1974–1998 | Environment Protection Principles | 1974 | 1992 | | | Environment Protection Principles 1974–1992 |
| | | | 1992 | | | Environment Protection Principles 1992–1998 |
| | Nature Conservation | 1992 | | | 1998 | Nature Conservation 1992–1998 |
| | Principles of Cooperation | 1974 | 1992 | | | Principles of Cooperation 1974–1992 |
| | | | 1992 | | 1998 | Principles of Cooperation 1992–1998 |
| | Regulations for All Sources of Marine Pollution | 1974 | 1992 | | | Regulations for All Sources of Marine Pollution 1974–1992 |
| | | | 1992 | | 1998 | Regulations for All Sources of Marine Pollution 1992–1998 |

**Table 2.2**
(continued)

| Regime | Regime component | Formation (Year) | Watershed (Year) | Second watershed (Year) | Endpoint of coding (Year) | Regime element |
|---|---|---|---|---|---|---|
| Barents Sea Fisheries Regime 1975–1998 | Norwegian-Russian Cooperation on Fisheries in the Barents Sea Region | 1975 | | | 1998 | Norwegian-Russian Cooperation on Fisheries in the Barents Sea Region 1975–1998 |
| Biodiversity Regime 1992–1998 | Convention on Biological Diversity 1992–1998 | 1992 | | | 1998 | Convention on Biological Diversity 1992–1998 |
| CITES Regime 1973–1998 (Trade in Endangered Species) | CITES Convention | 1973 | 1989 | | | CITES Convention 1973–1989 |
| | | | 1989 | | 1998 | CITES Convention 1989–1998 |
| | TRAFFIC Network on Monitoring and Compliance | 1978 | 1989 | | | TRAFFIC Network on Monitoring and Compliance 1978–1989 |
| | | | 1989 | | 1998 | TRAFFIC Network on Monitoring and Compliance 1989–1998 |

| | | | | | | |
|---|---|---|---|---|---|---|
| Climate Change Regime 1992–1998 | United Nations Framework Convention on Climate Change (UNFCCC) | 1992 | 1997 | | | UNFCCC 1992–1997 |
| | | | 1997 | | 1998 | UNFCCC 1997–1998 |
| | UNFCCC Financial Mechanism | 1992 | 1997 | | | UNFCCC Financial Mechanism 1992–1997 |
| | | | 1997 | | 1998 | UNFCCC Financial Mechanism 1997–1998 |
| | Kyoto Protocol to UNFCCC | 1997 | | | | Kyoto Protocol to UNFCCC 1997–1998 |
| Danube River Protection Regime 1985–1998 | Danube River Protection | 1985 | 1991 | | | Danube River Protection 1985–1991 |
| | | | 1991 | 1994 | | Danube River Protection 1991–1994 |
| | | | | 1994 | 1998 | Danube River Protection 1994–1998 |
| Desertification Regime 1994–1998 | United Nations Convention to Combat Desertification | 1994 | | | 1998 | United Nations Convention to Combat Desertification 1994–1998 |

**Table 2.2**
(continued)

| Regime | Regime component | Formation (Year) | Watershed (Year) | Second watershed (Year) | Endpoint of coding (Year) | Regime element |
|---|---|---|---|---|---|---|
| Great Lakes Management Regime 1972–1998 | Great Lakes Water Quality | 1972 | 1978 | | | Great Lakes Water Quality 1972–1978 |
| | | | 1978 | | 1998 | Great Lakes Water Quality 1978–1998 |
| | Great Lakes Water Quality | 1972 | 1978 | | | Great Lakes Water Quality 1972–1978 |
| | | | 1978 | | 1998 | Great Lakes Water Quality 1978–1998 |
| | Great Lakes Ecosystem Management | 1978 | | | 1998 | Great Lakes Ecosystem Management 1978–1998 |
| Hazardous Waste Regime 1989–1998 | Basel Convention | 1989 | 1995 | | | Basel Convention 1989–1995 |
| | | | 1995 | | 1998 | Basel Convention 1995–1998 |
| | Amendment to the Basel Convention | 1995 | | | 1998 | Amendment to the Basel Convention 1995–1998 |
| | OECD/EU/Lome IV Regulations | 1989 | 1995 | | | OECD/EU/Lome IV Regulations 1989–1995 |

| | | | | | |
|---|---|---|---|---|---|
| | | | 1995 | 1998 | OECD/EU/Lome IV Regulations 1995–1998 |
| | Bamako Convention | 1991 | 1995 | | Bamako Convention 1991–1995 |
| | Bamako/Waigani Conventions | | 1995 | 1998 | Bamako/Waigani Conventions 1995–1998 |
| IATTC Regime 1949–1998 (Interamerican Tropical Tuna Convention) | Conservation and Management of Tunas and Tuna-Like Fishes | 1949 | 1976 | | Conservation and Management of Tunas and Tuna-Like Fishes 1949–1976 |
| | | | 1976 | 1998 | Conservation and Management of Tunas and Tuna-Like Fishes 1976–1998 |
| | Conservation and Management of Dolphins | 1976 | | 1998 | Conservation and Management of Dolphins 1976–1998 |
| ICCAT Regime 1966–1998 (Conservation of Atlantic Tunas) | ICCAT Convention | 1966 | | 1998 | ICCAT Convention 1966–1998 |

**Table 2.2**
(continued)

| Regime | Regime component | Formation (Year) | Watershed (Year) | Second watershed (Year) | Endpoint of coding (Year) | Regime element |
|---|---|---|---|---|---|---|
| Regime for the International Regulation of Whaling 1948–1998 | International Convention for the Regulation of Whaling | 1946 | 1982 | | | Whaling Regime 1946–1982 |
| | | | 1982 | | 1998 | Whaling Regime 1982–1998 |
| London Convention Regime 1972–1998 | Wastes and Substances the Dumping of Which Is Prohibited | 1972 | 1991 | | | Wastes and Substances the Dumping of Which Is Prohibited 1972–1991 |
| | | | 1991 | | 1998 | Wastes and Substances the Dumping of Which Is Prohibited 1991–1998 |
| | Wastes and Substances Which, in Principle, May Be Dumped | 1972 | 1991 | | | Wastes and Substances Which, in Principle, May Be Dumped 1972–1991 |
| | | | 1991 | | 1998 | Wastes and Substances Which, in Principle, May Be Dumped 1991–1998 |

| | | | | | |
|---|---|---|---|---|---|
| | Regulation of Incineration at Sea | 1978 | 1991 | | Regulation of Incineration at Sea 1978–1991 |
| | | | 1991 | 1998 | Regulation of Incineration at Sea 1991–1998 |
| ECE Regime on Long-Range Transboundary Air Pollution 1979–1998 | LRTAP Convention | 1979 | 1982 | | LRTAP Convention 1979–1982 |
| | | | 1982 | 1998 | LRTAP Convention 1982–1998 |
| | First Sulphur Protocol | 1985 | | 1998 | First Sulphur Protocol 1985–1998 |
| | Second Sulphur Protocol | 1994 | | 1998 | Second Sulphur Protocol 1994–1998 |
| | $NO_x$ Protocol | 1988 | | 1998 | $NO_x$ Protocol 1988–1998 |
| | VOCs Protocol | 1991 | | 1998 | VOCs Protocol 1991–1998 |
| North Sea Regime 1972/1974–1998 | North Sea Conferences | 1984 | | 1998 | North Sea Conferences 1984–1998 |
| | OSCOM/PARCOM | 1972/1974 | 1984 | | OSCOM/PARCOM 1972/1974–1984 |
| | OSCOM/PARCOM/OSPAR | | 1984/1992 | 1998 | OSCOM/PARCOM/OSPAR 1984/1992–1998 |

**Table 2.2**
(continued)

| Regime | Regime component | Formation (Year) | Watershed (Year) | Second watershed (Year) | Endpoint of coding (Year) | Regime element |
|---|---|---|---|---|---|---|
| Oil Pollution Regime 1954–1998 | International Convention for the Prevention of Pollution of the Sea by Oil (Oilpol) | 1954 | 1978 | | | Oilpol 1954–1978 |
| | International Convention for the Prevention of Pollution from Ships (MARPOL) | | 1973/1978 | | 1998 | MARPOL 1973/1978–1998 |
| | Regional Memoranda of Understanding | 1982 | | | 1998 | Regional Memoranda of Understanding 1982–1998 |
| Regime for Protection of the Rhine Against Pollution 1963–1998 | Berne Convention | 1963 | | | 1998 | Berne Convention 1963–1998 |
| | Chemical Pollution Convention | 1976 | | | 1998 | Chemical Pollution Convention 1976–1998 |
| | Chloride Pollution Convention | 1976 | | | 1998 | Chloride Pollution Convention 1976–1998 |
| | Ecosystem/Salmon (Rhine Action Plan) | 1987 | | | 1998 | Ecosystem/Salmon (RAP) 1987–1998 |

| | | | | | | |
|---|---|---|---|---|---|---|
| Ramsar Regime on Wetlands 1971–1998 | Ramsar Convention on Wetlands | 1971 | 1987 | | | Ramsar Convention 1971–1987 |
| | | | 1987 | | 1998 | Ramsar Convention 1987–1998 |
| Regime for Protection of the Black Sea 1992–1998 | Bucharest Convention and Protocols for Protection of the Black Sea against Pollution | 1992 | | | 1998 | Bucharest Convention and Protocols for Protection of the Black Sea against Pollution 1992–1998 |
| | Black Sea Strategic Action Plan | 1996 | | | 1998 | Black Sea Strategic Action Plan 1996–1998 |
| South Pacific Fisheries Forum Agency Regime 1979–1998 | General Management of Fisheries in the South Pacific Region | 1979 | 1982 | | | General Management of Fisheries in the South Pacific Region 1979–1982 |
| | | | 1982 | 1995/1997 | | General Management of Fisheries in the South Pacific Region 1982–1995/1997 |
| | | | | 1995/1997 | 1998 | General Management of Fisheries in the South Pacific Region 1995/1997–1998 |

**Table 2.2**
(continued)

| Regime | Regime component | Formation (Year) | Watershed (Year) | Second watershed (Year) | Endpoint of coding (Year) | Regime element |
|---|---|---|---|---|---|---|
| | Compliance of Fisheries Management in the South Pacific Region | 1979 | 1982 | | | Compliance of Fisheries Management in the South Pacific Region 1979–1982 |
| | | | 1982 | 1995/1997 | | Compliance of Fisheries Management in the South Pacific Region 1982/1995–1997 |
| | | | | 1995/1997 | 1998 | Compliance of Fisheries Management in the South Pacific Region 1995/1997–1998 |
| Stratospheric Ozone Regime 1985–1998 | Vienna Convention | 1985 | 1990 | | | Vienna Convention 1985–1990 |
| | | | 1990 | | 1998 | Vienna Convention 1990–1998 |
| | Montreal Protocol | 1987 | 1990 | | | Montreal Protocol 1987–1990 |
| | | | 1990 | | 1998 | Montreal Protocol 1990–1998 |

| | | | | |
|---|---|---|---|---|
| | London Amendment | 1990 | 1998 | London Amendment 1990–1998 |
| | Copenhagen Amendment | 1992 | 1998 | Copenhagen Amendment 1992–1998 |
| | Multilateral Fund | 1990 | 1998 | Multilateral Fund 1990–1998 |
| Tropical Timber Trade Regime | International Tropical Timber Agreement | 1983 | 1998 | International Tropical Timber Agreement 1983–1998 |

handful of the nearly 100 individuals identified appeared more than once. Expert coders often found it difficult to assess the importance of individuals. Given the facts that most regimes have long lifespans and that most individuals are active for much shorter periods, this is understandable. Information on both nonstate actors and individuals is limited to the roles they play in political processes. With regard to important states, by contrast, the IRD contains information pertaining both to their records of compliance with regime rules and to the evolving nature of their knowledge bases.

## 2.4 Coding Procedures and the Use of the IRD

Before we turn to the applications of the IRD in chapters 3 to 5, several issues pertaining to coding procedures and data analysis merit attention. A database relying on a single coding for each of the regimes covered is vulnerable to criticisms regarding internal validity. It is quite possible that separate coders would respond in different ways to the questions included in the data protocol. We therefore asked two experts to code each regime independently. As result, the IRD contains double datasets for 21 of the 23 regimes included or, to put it in other terms, for 172 regime elements. Because we asked our expert coders to respond to questions only when they felt sufficiently knowledgeable to answer them with confidence, there are limits to the comparability of these double datasets. Nevertheless, comparing these datasets makes it possible to investigate issues of intercoder reliability in a systematic fashion.

Most of the coders we recruited are political scientists and international lawyers who share a general conception of regimes as legal arrangements created to address transboundary problems (Slaughter, Tulumello, and Wood 1998). These individuals are recognized in the community for their expertise on the regimes they coded. In a number of instances, however, one of the coders came from another discipline or from a nonacademic setting. Some officials of regime secretariats, members of national delegations to international negotiations, representatives of nongovernmental organizations, and independent consultants served as coding experts. To avoid misunderstandings on the part of coders coming from nonacademic settings, we amplified the descriptions of individual variables included in the data protocol. The trial runs also led to efforts to clarify key variables. The result was a high level of shared understanding among the coders regarding the meaning of questions, variables, and the nominal or ordinal scales included in the forms of our data protocol.

Needless to say, none of this rules out the possibility of bias on the part of our expert coders. The fact that a great many of them are lawyers or political scientists, for example, may incline them toward exaggerating the role of institutional arrangements. This may be especially true with respect to judgments about matters like goal attainment or problem solving. There is no such thing as a completely objective system of coding in research of this type; our work is certainly no exception to this proposition. Nonetheless, we approached this matter with care; we took a number of steps (e.g., engaging in discussions with the coders in cases where their answers diverged substantially) intended to minimize bias in the coding process. Under the circumstances, we have a reasonably high level of confidence in the results.

The data protocol is formatted as a questionnaire requesting both quantitative and qualitative answers to a large number of questions. Coding consisted of several steps, beginning with the negotiation of precoding agreements between the coders working on a specific regime and members of the IRD management team. Once the precoding agreement was in place, the management team was able to customize the protocol to fit the case at hand. At this point, the coders could begin to make their way through the protocol answering individual questions to the best of their ability.

As data on individual regimes arrived, the database team sought to clarify ambiguous answers and to reconcile responses where there appeared to be significant differences between the coders. Many differences resulted from misunderstandings or divergent interpretations concerning the intent of specific questions and were easily resolved. We took several additional steps to maximize the quality of data. Coders were allowed to indicate levels of (un)certainty regarding their answers to individual questions in the data protocol. In the aftermath of the coding exercise, we discussed these judgments of uncertainty with individual coders, along with other questions emerging from a check of the data they had delivered.

In some cases, the two experts coded certain variables dealing with regime attributes together before embarking on their individual coding efforts. Joint coding of variables describing a regime's principles, norms, and rules as well as its goals, programmatic activities, and administrative bodies evolved into a continuation of the precoding negotiations and helped to establish shared understandings between the two coders. Still, the two experts coded roughly 90 percent of the 136 forms included in the IRD codebook independently of one another. The management team (including student assistants) ultimately transformed the coders' answers into an electronically accessible database organized around each of the variables included in the data protocol.

Experts and members of the management team confronted several additional problems during the coding process. Given the scope of the IRD, it was clear that some coders would find it difficult or impossible to respond to all the queries included in the data protocol. In cases where characteristics of the data protocol itself caused the problem, the database team sought to eliminate—or at least alleviate—the problem through a dialogue with the coders. As mentioned, we also developed a rule asking coders to provide answers only to queries they could respond to with confidence. This has the effect of limiting the database with regard to certain matters (e.g., the scope and content of the rules included in some regimes and some matters of implementation like systems of implementation review). It also has consequences for the number of cases included in the database. The inability of coders to provide answers to some queries means that complete data are available for less than the 172 cases theoretically included in the database. In some cases, we also adopted a practice of asking coders to communicate with one another in responding to a specific query. The query dealing with the programmatic activities of regimes is a case in point. It is only fair to identify these issues explicitly; they certainly do constitute limitations of the data included in the IRD. By the same token, however, they actually strengthen the database with regard to the problem of internal validity. By excluding educated guesses on the part of the coders, for instance, we have limited the amount of unreliable data included in the IRD.

Users of the IRD can choose among several types of statistical analysis. The database treats every regime element coded by a case-study expert as a separate record. Since we have 172 regime elements, the number of individual records in the database is large. In the substantive chapters of this book, we rely mainly on uni- and bivariate analyses. But others will be able to use more advanced statistical methods in exploring their own research questions.

The IRD is stored as a set of files in MS Access. As a result, we developed all queries used as a basis for the empirical analyses reported in chapters 3–5 as select queries in MS Access. Because this is a well-known and widely available software package, analysts will find the IRD easy to access. The data themselves are loaded on the CD included in this book; members of the management team will be happy to engage in a dialogue with both academics and nonacademics who demonstrate a serious interest in using the IRD to engage in systematic research on international regimes.[1]

## 2.5 Conclusion

Described in these simple terms, the construction of the database appears to be a routine process moving steadily from stage to stage in a straightforward manner. But in actuality, the process was highly labor intensive and fraught with a steady stream of difficult choices. We know now that the construction of a computerized database on the scale reflected in the IRD is an enormous undertaking. We have sought to address all the issues arising in this process squarely and to make decisions in each case based on a careful weighing of the options. Any final assessment of the appropriateness of the decisions we have made must await the production and evaluation of scientific results on the part of users of the IRD.

Our efforts have led not only to the development of a foundation for the quantitative study of international regimes but also to a sharpening of many of the key concepts associated with regime analysis. Some may be tempted to conclude that the project has tried to address too many variables. Individual coders may be forgiven for arriving at such a conclusion as a result of the effort required to answer all the questions included in the data protocol. Even so, others may conclude that the data protocol omits or does an inadequate job of addressing some variables of interest to those seeking to understand the formation, consequences, and evolution of international regimes.

The next three chapters not only illustrate uses of the IRD but also demonstrate the links between the database and certain recurrent research themes. While the database provides excellent materials for a study of regime outputs, outcomes, and impacts, for instance, it does not do full justice to emerging questions relating to the legitimacy of international governance systems or to the interplay of distinct regimes in a social space in which the density of institutional arrangements is increasing. This project demonstrates that the creation of any tool for the systematic study of international regimes requires a major investment of time and energy. But we believe it will also convince students of international institutions, policymakers, funders, and members of the broader public that a systematic study of the contributions regimes make to the pursuit of global governance is worthwhile. The IRD is a tool for examining the causes of both success and failure in efforts to respond to the demand for international and transnational governance. It is already clear that both complete success and complete failure are exceptional. But we need to go beyond this overall conclusion. Using the IRD as a tool for analysis, it is realistic to expect that we can arrive at more specific conclusions about such matters as the roles of monitoring systems, compliance mechanisms, funding arrangements, systems of implementation review, and so forth as determinants of the effectiveness of international regimes.

# 3

# Sources of Compliance with International Regimes

## 3.1 Introduction

Effective regulation depends on a mutual belief in cooperation.[1] Such a belief requires in turn a high compliance rate for any given regulation. High compliance rates, however, are assumed to depend on two conditions that seldom prevail outside the institutionalized framework of the developed nation-state: an established monopoly of legitimate force and a national identity that produces consent on the part of those who are targets of regulations, even if they consider the rules in question inconvenient. In this sense, it seems fair to ask whether compliance is the Achilles' heel of international regulation (see Werksmann 1996, xvi; Young 1999b, chap. 4).

Those who doubt the effectiveness of international institutions argue that successful regulation requires centralized coercion administered by an agent in possession of superior resources. Otherwise, compliance with inconvenient commitments becomes a question of opportunism—an observation that runs counter to the idea that norms and rules matter. On this account, high compliance rates with international regulations are impossible, as long as they require signatories to a treaty to do things they would otherwise prefer not to do. When high compliance rates do occur, they are attributed to shallow treaties that involve little "depth of cooperation" (Downs, Rocke, and Barsoon 1996).

Those who believe in the effectiveness of international regulations, by contrast, start with the observation that international norms and rules are often complied with to a remarkably high degree. "Almost all nations observe almost all principles of international law and all of their obligations almost all of the time" is Louis Henkin's (1979, 47) frequently cited conclusion. Scholars who endeavor to understand how international regulations work in the environmental field have revived this observation.[2] According to their findings, it is not so much coercion by a superior power but good management and institutionalized incentive mechanisms that lead to satisfactory levels of compliance. In the words of Abram Chayes and Antonia Handler Chayes (1993, 205), "Enforcement through these interacting measures of assistance and persuasion is less costly and intrusive and is certainly less dramatic than coercive sanctions, the easy and usual policy elixir for non-compliance."

These observations regarding effective regulation beyond the nation-state encapsulate two separate processes that can generate adequate levels of compliance (Zürn 2004). One involves a focus on *softer paths* to compliance that are based on rational consent, including capacity building, legitimacy building, and the voluntary internalization of law. The other posits that to the extent that coercive sanctions are used as a legitimate means of generating compliance, they need not be applied only in a hierarchical context but can also be used in an institutionalized *horizontal setting*.[3] In this view, compliance with regulations varies with the availability of information, institutionalized horizontal coercion, shaming, and adjudication, among other things.

In this chapter, we explore horizontal mechanisms for inducing compliance, setting forth a range of theoretical arguments and comparing them with the empirical data contained in the International Regimes Database (IRD). Section 2 discusses the dependent variable as treated in the IRD. In section 3, we explore four different approaches to accounting for compliance with international regulations. In each case, we derive explicit hypotheses, compare them with evidence drawn from the IRD, and discuss the significance of our findings. The concluding section addresses the significance of these findings for the debate about compliance described in the preceding paragraphs.

## 3.2 Compliance in the International Regimes Database

Compliance differs from other topics dealing with the consequences of regulations, including their implementation and effectiveness. Of course, there are many points of contact, overlaps, and links between these phenomena. But the focus of research on implementation is an analysis of differences between legislative requirements on paper and the forms they take in practice (Victor, Raustiala, and Skolnikoff 1998, 4). The principal target of effectiveness research is the capacity of political institutions to solve commonly perceived problems (Young 1999a).

Compliance research is distinct; it examines the extent to which subjects comply with rules addressed to them. Thus, "compliance can be said to occur when the actual behavior of a given subject conforms to prescribed behavior, and non-compliance or violation occurs when actual behavior departs significantly from prescribed behavior" (Young 1979, 3). Moreover, "compliance is a noun that denotes a particular type of *behavior, action or policy* within a specific regulatory or situational context." (Simmons 2000, 1; emphasis in original). It does not refer to the *willingness* of the actors to comply. The object of our empirical study, therefore, is the actions or behavior of subjects, rather than their attitudes or motives.

The fact that all rules are more or less ambiguous necessitates application and interpretation and thus makes it hard to assess objectively whether or not compliance is taking place. What is more, laws and norms are not constant over time; they are subject to changing interpretations and, in judicial settings, to new case law that interprets and often changes the meaning of statutory law (Dworkin 1986). Nevertheless, it is possible to assess compliance from an external perspective by making systematic use of indicators of internal estimates of compliance. As Simmons (2000, 24) puts it, "The point is to compile objective evidence of subjective socially-based interpretations of behavior."

In this chapter, we focus on the behavioral aspect of compliance. To be sure, compliance is a two-dimensional phenomenon consisting of a substantive dimension—the relationship between obligations and actual behavior—and a procedural dimension—the treatment of accusations of noncompliance (Weiss and Jacobson 1998, 4). At least at the margins, the concepts of compliance and noncompliance are perpetually contested; their meaning evolves over time once a rule moves from paper to practice. Charges of noncompliance can and often do arise from the ambiguity of rules, even when the subjects have no incentive to cheat or to challenge the validity of the rules. The assumption, in such cases, is that compliance is not problematic once differences about the correct interpretation of the rule are ironed out. We do not ignore these complexities. Nonetheless, we focus here on the substantive dimension of compliance as our dependent variable.

In the IRD, the questions about compliance in the substantive sense are straightforward. Does the behavior of important actors generally conform to the provisions of the regime? Did the regime exert a causal influence on these developments? Coders were asked to distinguish important nations, all members, and nonstate actors in order to provide a general judgment about compliance as well as to direct attention to the behavior of specific actors deemed important to the success of the regime at hand. The level of compliance for each category of subjects can vary over a five-point scale ranging from "behavior exceeds requirements" to "behavior does not conform at all." Intermediate levels include "behavior meets requirements," "behavior conforms with some requirements but not all," and "behavior conforms some (but not all) of the time and/or to some degree but not completely."

In addition, we asked coders to assess the causal impact of the regime on behavior by differentiating "little or no causal impact," "modest causal influence," and "large causal influence." Coders were able to specify that the causal impact was "positive" or "negative." Of course, they had the option of saying "don't know" when they were unable to make a causal judgment.

As table 3.1 shows, compliance levels recorded in the IRD generally confirm Louis Henkin's hypothesis, especially if it is stated less starkly. In a more modest fashion, we can conclude that the majority of member states comply with the majority of international environmental rules most of the time.[4] With respect to the variable compliance, the IRD contains 130 elements coded for the degree to which the behavior of all members conformed with the provisions of a regime. These data encompass 22 of the 23 regimes included in the IRD. For 80 of these elements, behavior meets regime requirements (62) or even exceeds them (18). Thus, more than 60 percent of the data collected for these 130 coded elements reflect compliance behavior that lives up to the requirements of regime rules.

**Table 3.1**
Conformity with regime provisions and causal influence of regimes

| | | CONFORMITY_CAUSAL | | | | | | | | | |
|---|---|---|---|---|---|---|---|---|---|---|---|
| | | All cases | | 1 | | 2 | | 3 | | 4 | |
| CONFORMITY | 1 | 18 | 13.8% | 0 | 0.0% | 0 | 0.0% | 17 | 21.3% | 1 | 10.0% |
| | 2 | 62 | 47.7% | 3 | 37.5% | 20 | 62.5% | 35 | 43.8% | 4 | 40.0% |
| | 3 | 34 | 26.2% | 3 | 37.5% | 8 | 25.0% | 21 | 26.3% | 2 | 20.0% |
| | 4 | 13 | 10.0% | 1 | 12.5% | 2 | 6.3% | 7 | 8.8% | 3 | 30.0% |
| | 5 | 3 | 2.3% | 1 | 12.5% | 2 | 6.3% | 0 | 0.0% | 0 | 0.0% |
| | Total | 130 | 100.0% | 8 | 100.0% | 32 | 100.0% | 80 | 100.0% | 10 | 100.0% |

CONFORMITY: 1 = behavior exceeds regime requirements, 2 = behavior meets regime requirements, 3 = behavior conforms with some requirements but not all, 4 = behavior conforms some (but not all) of the time and/or to some degree but not completely, 5 = behavior does not conform at all
CONFORMITY_CAUSAL: 1 = little or no causal impact, 2 = modest causal influence, 3 = large causal influence, 4 = negative causal influence

At the same time, the data show that international environmental regimes frequently encounter compliance problems. In almost 40 percent of the records, some rules elicit insufficient compliance, at least some of the time. Complete noncompliance occurs rarely—only in 2.3 percent of the records. Even in cases of insufficient compliance, progress is possible if compliance problems are tackled institutionally. In cases where watersheds occur, levels of compliance generally improve following the watershed. Compliance improved in 15.4 percent of the cases after a watershed; deterioration in compliance occurred only in 2.5 percent of the cases, despite the fact that rules and regulations generally become more demanding or more encompassing in the aftermath of a watershed.

Even more remarkable is the fact that coders ascribed a significant causal role to the regimes included in the IRD. Coders saw a large (61.5 percent) or at least "modest causal influence" (24.5 percent) in most of the cases. They failed to find causal impact in only 6 percent of the cases. Of course, these assessments are interpretative in nature and may reflect a *déformation professionnelle* on the part of regime analysts and regime practitioners. Given the impressive numbers, however, it seems safe to assume that international environmental regimes do have a causal impact on compliant behavior. This finding challenges the views of Downs, Rocke, and Barsoom (1996) and others who argue that high compliance rates are attributable to the shallowness of regimes. Where compliance rates actually exceed regime requirements, all but one of the coders (94.5 percent) pointed to a large causal inference. This finding implies that there are mechanisms at work in this setting that produce overcompliance. Neither the legitimacy of the rules themselves nor formal sanctioning of rule breakers provides an obvious explanation for overcompliance.

Some brief observations about specific regimes can help to flesh out the overall picture. The behavior of relevant actors has exceeded regime requirements over time in a few issue areas. The Antarctic Treaty System, the Baltic Sea Regime, the Great Lakes Water Quality Regime, the regime for the protection of the Rhine River against pollution, and the stratospheric ozone regime are all cases in point. Compliance problems plagued some of these regimes at the outset. But in each case, the development of the regime has overcome them. Regime development over time is thus an important issue for understanding compliance with international regulations. In only one case, the Barents Sea Fisheries Regime, has the regime started out with good compliance and maintained this record across issues and over time.

It is also important to note that in all the cases of good or exceptional compliance, the environmental problem has changed for the better. What is more, the coders judged that the institutional features of the regime played a large causal role in these processes. In short, experts and practitioners in the field do not believe that positive results in these environmental issue areas could have taken place in the absence of these international regimes.

Examples of less successful regimes include the biodiversity regime, the desertification regime, the Ramsar regime for wetlands, and the tropical timber regime. In each of these cases, at least some of the regime's requirements are met only partially. Although compliance problems in these cases are not minor, they need not threaten the existence of a regime or negatively affect the willingness of other members to comply over time. In fact, compliance problems can foster regime development.

From a dynamic perspective, we can see a trend toward launching efforts to tackle compliance problems, a development that often yields positive results in terms of improvements in compliance. At present, for example, the behavior of members of the Ramsar regime on wetlands does not meet regime requirements fully. But there are constant efforts to improve the implementation of the regime at the global level. The parties regularly adopt multiyear strategic plans. And an assessment of national reports received from 106 Contracting Parties pertaining to the implementation of the convention's strategic plan over the three-year period 1996 to 1999 concludes that good progress has been made in the implementation of the convention, despite the existence of continuing problems.[5]

There are as well some regimes where coders disagreed about levels of compliance. The Baltic Sea regime for the period before 1990, for instance, is one of the rare cases in which the academic expert and the practitioner disagreed strongly about the level of overall compliance. In this case, a lack of data regarding national compliance before 1989 may have contributed to the problem of assessing compliance. But this case also makes it clear that external analysts and members of regime secretariats can have different judgments about a regime's achievements. It would be presumptuous of us to treat one assessment as more authoritative than the other. Beyond this, coders observed that individual states encountered problems in implementing HELCOM Recommendations effectively, particularly prior to the early 1990s. But more recent evidence of reductions in the loads of various pollutants indicates that levels of compliance rose during the 1990s (Roginko 1998, 617).

Do states that play particularly important roles regarding environmental problems behave differently than others? Each coder of an IRD case identified four to seven important states for the case at hand. Not surprisingly, the United States is listed as an important state for 119 out of 130 coded elements. A remarkable finding regarding these important states is that their rates of compliance with the provisions of international regimes surpass those of other states. In 84 percent of the records, their behavior meets or exceeds regime requirements, compared to 62 percent for compliance of all members. Moreover, this finding is not due simply to superior capacity on the part of important states or to the logic of the lowest common denominator, according to which rich and powerful states exceed the behavioral requirements of international regimes in any case. In the majority of the cases included in the IRD, coders have judged that institutional features have a large causal impact on compliance.

What can we say about specific states in this context? The United States performs slightly better than the average for all important nations. The coders assigned a large causal role to the regime in inducing compliance on the part of the United States in 75 percent of the cases. The performance of other important western states is similar and sometimes even more striking. Germany, for instance, exceeds or meets regime requirements in 83 percent of the cases. Similar numbers apply to Denmark and Great Britain.

The performance of Russia/Former Soviet Union is more ambiguous. With the exception of the Barents Sea Fisheries Regime, Russia fulfills regime requirements only partially. In some instances, however, Russian compliance has improved over time. More or less severe problems regarding Russian compliance with the whaling regime during its early period (1946–1982), for instance, have been alleviated more recently. A similar pattern characterizes Russian compliance with the requirements of CCAMLR as a component of the Antarctic Treaty System.

Overall, we can summarize the evidence regarding compliance as the dependent variable as follows. Member states meet or surpass regime requirements in the majority of cases. Important member states do even better than other states regarding compliance with environmental regulations. Nevertheless, compliance is not automatic. It is deficient in a significant number of cases. Yet even in cases of deficient compliance, there is a trend toward improvements over time in rates of compliance.

## 3.3 Determinants of Compliance

Evidence drawn from the regimes database demonstrates that compliance with rules and regulations does not require the existence of a political hierarchy and a legitimate monopoly on the use of force and thus a national context. Rather, mechanisms like legitimacy, legalization, responsiveness, and the use of horizontal coercion appear to determine

Table 3.2
Sources of noncompliance

| | Rule is not challenged | Rule is challenged |
|---|---|---|
| **Voluntary noncompliance** | Cheating | Norm considered wrong |
| **Involuntary noncompliance** | Ambiguity/impreciseness of a prescription | Lack of capacity to implement/inadvertent noncompliance |

levels of compliance. As a result, we need to learn more about the nature and operation of compliance mechanisms in horizontal contexts. In this section, we categorize these mechanisms by focusing on the *sources of noncompliance* they address. We differentiate four sources of noncompliance, by examining first the extent to which noncompliance is voluntary or involuntary and second whether noncompliance amounts to a substantial challenge to the rule in question or whether it essentially involves a "technical" problem (table 3.2).[6]

*Cheating* takes place when actors with a clear understanding of a rule choose to violate it to their own advantage. Behavior of this type is generally secret, and the advantage is usually opportunistic in the sense that it depends on the persistence of the relevant obligation as such. The use of hidden subsidies to provide domestic industries with an advantageous position at the international level would constitute cheating. Noncompliance due to the *ambiguity of a rule* is different because it does not require secret activities and does not even necessarily benefit the party charged with noncompliance. Thus, expenditures motivated by the desire to strengthen research and development might be interpreted as rule-breaking subsidies, whether or not they are effective. When a *norm or rule is considered wrong*, parties may voluntarily and openly disregard it, even if it is formulated precisely. In these cases, transgressors point to the wrongness of the rule to justify noncompliance. In the case of the climate regime, for example, some governments have denounced openly the obligation to reduce $CO_2$ emissions. Civil disobedience is a particularly clearcut example of this type of noncompliance. Open but involuntary violations occur when *rules turn out to be impractical.* In these cases, there is no debate about the correct interpretation or the validity of the

**Table 3.3**
Theoretical perspectives on compliance

| | **Rule is not challenged** | **Rule is challenged** |
|---|---|---|
| **Voluntary noncompliance** | Incentives | Legitimacy |
| **Involuntary noncompliance** | Legalization | Responsiveness |

rule, but parties emphasize practical limitations in fulfilling it. Less developed states often encounter such difficulties when they sign ambitious international environmental treaties.

Each of the four sources of noncompliance corresponds to a specific theoretical perspective on compliance.[7] In this discussion, we label the four perspectives "incentives," "legalization," "legitimacy," and "responsiveness" (table 3.3).

Each perspective also calls for a certain method of curing noncompliance. They are not full-fledged theories of compliance. But they point to sets of variables that can help us to respond effectively to compliance problems. The four perspectives overlap in some ways. For instance, none of the perspectives denies the importance of monitoring compliance. Even so, they attach different weights to the role of key variables. Although these perspectives do not flow directly from broader theories of international relations, there are clear links to these theories. The theoretical perspective labeled "incentives" connects to the theory of international relations often called rational institutionalism. Other theories of international relations, such as liberalism and social constructivism, contribute to both the legalization perspective and the legitimacy perspective. The responsiveness perspective draws on other theories of international relations such as theories emphasizing discourses and communication. The four perspectives have affinities for distinct theories of law as well. In sum, each of the four theoretical perspectives on compliance involves a problem-driven process of theorizing, drawing on and combining elements of different analytic traditions in an effort to improve our understanding of the performance of international regimes.

### 3.3.1 Incentives against Cheating

The perspective labeled "incentives" treats cheating as the major problem and calls for the use of positive and negative sanctions (rewards and punishments) as the solution. It is based on the assumption that we are dealing with rational, unitary actors who comply with rules when and only when comparisons of costs and benefits produce positive results. But note that costs and benefits need not be limited to narrow economic considerations; they can include other factors like power or prestige. Two implications for compliance flow from this perspective. Each of the participating actors has an incentive to enjoy the benefits of cooperation without paying the costs or, in other words, to act as a free rider. In addition, there is a perpetual concern that cheating on the part of some actors will change the cost-benefit calculations of all participants, resulting in a breakdown of cooperation with negative consequences for all concerned.

This perspective, often labeled *rational institutionalism*, treats high levels of compliance with international rules as a possible outcome. It emphasizes that the participants' interests and motives regarding cooperation are mixed in most cases. Quite apart from an interest in the maximization of individual gains in the short run, considerations relating to longer-run consequences may lead actors to exercise caution and to behave in a way that does not endanger cooperative outcomes. In this sense, compliance can arise from a belief on the part of individual actors that cooperative behavior promotes their self-interest (Hurd 1999, 385). On this account, rule implementation is itself a process of bargaining of a type we may call compliance bargaining (Jönsson and Tallberg 1998). The incentive perspective thus differs in two important ways from realist thinking, which restricts effective regulation to systems featuring hierarchical relationships. Coercion in the form of threats and punishments is not necessary to achieve cooperation in all cases. While some successful institutions do require coercion to work, others do not rely on threats and punishments. What are often described as coordination problems in contrast to collaboration problems, for instance, do not produce situations in which individual participants experience an incentive to cheat.[8]

In addition, although all rules and regulations that generate incentives to cheat must be underpinned by threats and punishments to ensure their effectiveness and durability, it is sometimes possible to meet this requirement in the absence of hierarchical arrangements. In some situations, for example, the adoption of a strategy of tit for tat allows for the evolution of cooperation under anarchy (Axelrod 1984; Keohane 1984). In effect, horizontal coercion can function as a substitute for hierarchy or a legitimate monopoly of force.

In such settings, sustainable cooperation requires additional conditions that can make horizontal coercion effective. Numerous analysts have contributed to the identification of these conditions. Verification procedures need to be reliable and affordable. Verifying emissions of pollutants into a river, for example, is easier than verifying compliance with the rules of a high seas fishery regime. As a number of analysts have noted, institutional design often plays a role in determining how hard it is to verify compliance (Mitchell 1994b; Wettestad 1999). In addition, the sanctioning of violators should not be too costly for the parties imposing sanctions. Military sanctions, for instance, are almost always extremely costly, whereas sanctions in the form of retaliatory tariffs may actually prove beneficial. In most cases, the exercise of care in matching institutional design to the underlying problem structure can minimize the costs of sanctioning. In sum, the perspective labeled "incentives" links compliance to features of the institutional setting and characteristics of the problem structure that facilitate the verification of compliance and allow for the operation of horizontal enforcement mechanisms.

What can we learn about these theoretical expectations from an examination of the data included in the IRD? The most relevant form here is RA (in other words, regime attribute) 45: "Are there reporting procedures requiring the submission of information by individual members pertaining to regime implementation?"

Table 3.4 indicates that institutionalizing the verification of compliance is less important than rational institutionalism anticipates. When regimes include "reporting procedures requiring the submission of information by individual members pertaining to regime implementation," the level of compliance remains unchanged or declines insignificantly. In fact, the cell for "no reporting procedure/excessive compliance" includes a significantly higher share of the records than a normal distribution would predict. At first glance, this result is counterintuitive. The most plausible explanation is that in some cases verification of compliance is so easy and so transparent that there is no need for compliance mechanisms. Overall, the mere presence or absence of institutional reporting

Table 3.4
Compliance reporting

| | | COMPLIANCE_REPORTING | | | | | | | | | |
|---|---|---|---|---|---|---|---|---|---|---|---|
| | | All cases | | 0 = Not applicable | | 1 = No | | 2 = Yes | | 3 = Don't know | |
| CONFORMITY | 1 | 16 | 12.7% | 0 | 0.0% | 5 | 23.8% | 11 | 10.8% | 0 | 0.0% |
| | 2 | 60 | 47.6% | 1 | 100.0% | 9 | 42.9% | 48 | 47.1% | 2 | 100.0% |
| | 3 | 34 | 27.0% | 0 | 0.0% | 3 | 14.3% | 31 | 30.4% | 0 | 0.0% |
| | 4 | 13 | 10.3% | 0 | 0.0% | 2 | 9.5% | 11 | 10.8% | 0 | 0.0% |
| | 5 | 3 | 2.4% | 0 | 0.0% | 2 | 9.5% | 1 | 1.0% | 0 | 0.0% |
| | **Total** | **126** | **100.0%** | **1** | **100.0%** | **21** | **100.0%** | **102** | **100.0%** | **2** | **100.0%** |
| | | | | | | | | | | | |
| CONFORMITY_CAUSAL | 1 | 8 | 6.3% | 0 | 0.0% | 2 | 9.5% | 6 | 5.9% | 0 | 0.0% |
| | 2 | 32 | 25.4% | 1 | 100.0% | 4 | 19.0% | 25 | 24.5% | 2 | 100.0% |
| | 3 | 77 | 61.1% | 0 | 0.0% | 14 | 66.7% | 63 | 61.8% | 0 | 0.0% |
| | 4 | 0 | 0.0% | 0 | 0.0% | 0 | 0.0% | 0 | 0.0% | 0 | 0.0% |
| | 5 | 9 | 7.1% | 0 | 0.0% | 1 | 4.8% | 8 | 7.8% | 0 | 0.0% |
| | **Total** | **126** | **100.0%** | **1** | **100.0%** | **21** | **100%** | **102** | **100.0%** | **2** | **100.0%** |

CONFORMITY: 1 = behavior exceeds regime requirements, 2 = behavior meets regime requirements, 3 = behavior conforms with some requirements but not all, 4 = behavior conforms some (but not all) of the time and/or to some degree but not completely, 5 = behavior does not conform at all
CONFORMITY_CAUSAL: 1 = little or no causal impact, 2 = modest causal influence, 3 = large causal influence, 4 = negative causal influence, 5 = don't know

procedures in international environmental regimes is less important than generally expected.

A closer examination of more specific "procedures for reviewing implementation" (table 3.5) shows that soft forms of institutionalized verification are counterproductive, whereas strong "procedures for reviewing implementation" do increase compliance with regime provisions. While "information gathering for broad assessment" and "information gathering from third parties on implementation by other parties" have no impact—or even a negative impact—on compliance with treaty

Table 3.5
Implementation review (RA 46)

| | | Reviews | | | | | | | | | | | | | | | |
|---|---|---|---|---|---|---|---|---|---|---|---|---|---|---|---|---|---|
| | | All cases | | No entry | | 0 | | 1 | | 2 | | 3 | | 4 | | 5 | |
| CONFORMITY | 1 | 57 | 13.7% | 0 | 0.0% | 0 | 0.0% | 0 | 0.0% | 11 | 16.7% | 3 | 8.3% | 15 | 21.4% | 10 | 14.9% |
| | 2 | 179 | 42.9% | 1 | 100.0% | 4 | 80.0% | 12 | 70.6% | 26 | 39.4% | 9 | 25.0% | 24 | 34.3% | 30 | 44.8% |
| | 3 | 136 | 32.6% | 0 | 0.0% | 0 | 0.0% | 2 | 11.8% | 26 | 39.4% | 21 | 58.3% | 21 | 30.0% | 22 | 32.8% |
| | 4 | 38 | 9.1% | 0 | 0.0% | 1 | 20.0% | 2 | 11.8% | 1 | 1.5% | 3 | 8.3% | 10 | 14.3% | 5 | 7.5% |
| | 5 | 7 | 1.7% | 0 | 0.0% | 0 | 0.0% | 1 | 5.9% | 2 | 3.0% | 0 | 0.0% | 0 | 0.0% | 0 | 0.0% |
| Total | | 417 | 100.0% | 1 | 100.0% | 5 | 100.0% | 17 | 100.0% | 66 | 100.0% | 36 | 100.0% | 70 | 100.0% | 67 | 100.0% |

| | | | | 6 | | 7 | | 8 | | 9 | | 10 | | 11 | | 13 | |
|---|---|---|---|---|---|---|---|---|---|---|---|---|---|---|---|---|---|
| CONFORMITY | 1 | | | 0 | 0.0% | 4 | 10.3% | 4 | 19.0% | 4 | 17.4% | 4 | 33.3% | 2 | 5.9% | 0 | 0.0% |
| | 2 | | | 13 | 56.5% | 9 | 23.1% | 10 | 47.6% | 8 | 34.8% | 6 | 50.0% | 24 | 70.6% | 3 | 100.0% |
| | 3 | | | 8 | 34.8% | 19 | 48.7% | 5 | 23.8% | 7 | 30.4% | 0 | 0.0% | 5 | 14.7% | 0 | 0.0% |
| | 4 | | | 1 | 4.3% | 5 | 12.8% | 2 | 9.5% | 3 | 13.0% | 2 | 16.7% | 3 | 8.8% | 0 | 0.0% |
| | 5 | | | 1 | 4.3% | 2 | 5.1% | 0 | 0.0% | 1 | 4.3% | 0 | 0.0% | 0 | 0.0% | 0 | 0.0% |
| Total | | | | 23 | 100.0% | 39 | 100.0% | 21 | 100.0% | 23 | 100.0% | 12 | 100.0% | 34 | 100.0% | 3 | 100.0% |

REVIEW: 0 = not applicable, 1 = no explicit review procedures, 2 = information gathering for broad assessment without evaluating performance/compliance of individual parties, 3 = information from third parties on implementation by other parties, 4 = information gathering for assessment of performance and compliance of individual parties, 5 = review and broad assessment of the regime by the supreme decision-making body, 6 = review and broad assessment of the regime by bodies delegated by parties to make decisions or recommendations, 7 = review of member performance/compliance by the supreme decision-making body, 8 = review of member performance/compliance by bodies delegated by parties to make decisions or recommendations, 9 = recommendation/implementation of responses to inadequate performance by the supreme decision-making body, 10 = recommendation/implementation of responses to inadequate performance by bodies delegated by parties, 11 = on-site inspections to verify compliance, 13 = don't know

CONFORMITY: 1 = behavior exceeds regime requirements, 2 = behavior meets regime requirements, 3 = behavior conforms with some requirements but not all, 4 = behavior conforms some (but not all) of the time and/or to some degree but not completely, 5 = behavior does not conform at all provisions, the stronger forms of verification are associated positively with compliance. This is especially true for specific values of this variable, such as "recommendation/implementation of responses to inadequate performance by bodies delegated by parties" and "on-site inspections to verify compliance." The rates of behavior that fulfill or exceed requirements in these cases are 83.3 and 76.4 percent respectively (compared to an overall average of 56.6 percent for this variable)

Findings regarding the causal role institutionalized verification plays in behavioral terms support the thrust of these conclusions about the effects of verification on compliance. With the exception of a few particularly strong mechanisms, the role of verification regarding compliance appears to be less important than rational institutionalism predicts. The implication is clear: if you plan to institutionalize procedures to verify rule compliance, choose strong mechanisms.

The evidence concerning the role of institutionalized forms of horizontal enforcement is more impressive. It is both significant and in line with theoretical expectations (table 3.6).

The IRD asks coders to answer this question: "What formal compliance mechanisms are provided for in the regime's constitutive provisions to achieve compliance?" An examination of the data regarding those compliance mechanisms providing an opportunity for real enforcement (e.g., "suspension of membership rights," "exclusion from membership," "imposition of military punishments," and "imposition of financial and economic punishments") demonstrates that the presence of such mechanisms increases both compliance rates and the causal role regimes play in promoting compliance.

To summarize, the determinants of compliance identified by the theoretical perspective labeled "incentives" are important. Whereas institutionalized enforcement mechanisms have exactly the effects predicted by the theory, however, verification procedures make a difference only when these procedures take a strong form (e.g., on-site inspections). This observation points to the possibility that the problem with verification procedures may be a function of the situation structure in the issue area. This would be compatible with a consideration of incentives. If verification occurs "naturally" so that there is no need for institutionalized procedures, compliance rates are good. But when verification requires the use of institutionalized procedures, soft mechanisms do not suffice; strong mechanisms lead to improvements in compliance.

### 3.3.2 Legalization against Cheating

This perspective sees inconsistencies in rule development and application along with the abuse of ambiguities as the major sources of noncompliance. The solution is a process of legalization that incorporates the regulations in question as deeply as possible into existing rule-of-law systems. In addition, this perspective stresses the "preciseness" of rules and norms together with the importance of secondary rules that help to settle disputes over the content and the application of the rules and norms themselves.[9]

The fundamental assumption embedded in this perspective is that a legal system is more legitimate than a specific rule or regulation, a proposition implying that subjects may comply with a specific, lawlike rule because it is part of a larger legal system regarded as legitimate. This line of reasoning applies not only to national legal systems (Dworkin 1991), but also to the role of law in international society (Hurrell 1993). Subjects may comply with international rules, even when they run counter to self-interest more narrowly defined, because they regard them as part of an encompassing normative structure. Blatant, unjustified transgression of rules and regulations thus implies a general repudiation of the normative basis for international cooperation. In this sense, legal rules possess a compliance pull of their own (Franck 1990). It follows that the more a rule is integral to a legal system or, in other words, the more an international institution is legalized, the more likely compliance becomes.

The question that arises in this context is what makes decision makers perceive a rule as lawful. Here, we treat two sets of features as central to the process of legalization. We call one juridification and the other internalization (Zürn and Wolf 1999). *Juridification* refers to the processes that ensure that rules and regulations fulfill criteria like clarity, pertinence, stringency, adaptability, and a high degree of consistency both intrinsically and in relation to other laws. Abbott et al. (2000, 401) identify three elements of these processes:[10] "obligation" in the sense that states or other actors are bound by a rule or commitment; "precision" meaning that the rules define the conduct they require, authorize, or

**Table 3.6**
Compliance mechanisms (RA 47)

| | | FORMAL COMPLIANCE MECHANISMS | | | | | | | | | | | |
|---|---|---|---|---|---|---|---|---|---|---|---|---|---|
| | | All cases | | Sum of 0–2; 7–11* | | 3 | | 4 | | 5 | | 6 | |
| CONFORMITY | 1 | 32 | 17.2% | 29 | 17.2% | 3 | 42.9% | 0 | 0.0% | 0 | 0.0% | 0 | 0.0% |
| | 2 | 90 | 48.4% | 79 | 46.7% | 3 | 42.9% | 2 | 100.0% | 1 | 100.0% | 5 | 71.4% |
| | 3 | 45 | 24.2% | 43 | 25.4% | 1 | 14.3% | 0 | 0.0% | 0 | 0.0% | 1 | 14.3% |
| | 4 | 16 | 8.6% | 15 | 8.9% | 0 | 0.0% | 0 | 0.0% | 0 | 0.0% | 1 | 14.3% |
| | 5 | 3 | 1.6% | 3 | 1.8% | 0 | 0.0% | 0 | 0.0% | 0 | 0.0% | 0 | 0.0% |
| **Total** | | **186** | **100.0%** | **169** | **100.0%** | **7** | **100.0%** | **2** | **100.0%** | **1** | **100.0%** | **7** | **100.0%** |
| | | All cases | | Sum of 0–2; 7–11* | | 3 | | 4 | | 5 | | 6 | |
| CONFORMITY_CAUSAL | 1 | 9 | 4.8% | 9 | 5.3% | 0 | 0.0% | 0 | 0.0% | 0 | 0.0% | 0 | 0.0% |
| | 2 | 47 | 25.3% | 42 | 24.9% | 5 | 71.4% | 1 | 50.0% | 0 | 0.0% | 2 | 28.6% |
| | 3 | 20 | 10.8% | 108 | 63.9% | 2 | 28.6% | 1 | 50.0% | 1 | 100.0% | 5 | 71.4% |
| | 4 | 0 | 0.0% | 0 | 0.0% | 0 | 0.0% | 0 | 0.0% | 0 | 0.0% | 0 | 0.0% |
| | 5 | 13 | 7.0% | 10 | 5.9% | 0 | 0.0% | 0 | 0.0% | 0 | 0.0% | 0 | 0.0% |
| **Total** | | **89** | **47.8%** | **169** | **100.0%** | **7** | **100.0%** | **2** | **100.0%** | **1** | **100.0%** | **7** | **100.0%** |

FORMAL COMPLIANCE MECHANISMS: 0 = not applicable, 1 = no compliance mechanisms, 2 = issuance of notice of violation, 3 = suspension of membership rights, 4 = exclusion from membership, 5 = imposition of military punishments, 6 = imposition of financial/economic punishments, 7 = support for capacity building to enhance compliance, 8 = granting of a transition period to active compliance, 9 = dissolution of linkages, 10 = additional compliance mechanisms, 11 = don't know

CONFORMITY: 1 = behavior exceeds regime requirements, 2 = behavior meets regime requirements, 3 = behavior conforms with some requirements but not all, 4 = behavior conforms some (but not all) of the time and/or to some degree but not completely, 5 = behavior does not conform at all

CONFORMITY_CAUSAL: 1 = little or no causal impact, 2 = modest causal influence, 3 = large causal influence, 4 = negative causal influence, 5 = don't know

proscribe accurately and unambiguously; and "delegation" in the sense that authority is granted to third parties to take actions to implement the rules, including interpretation, application, dispute settlement, and further rule making. For our purposes, the element of obligation is a little too close to our dependent variable (compliance). But the other two elements point directly to the problem of ambiguity. The major response to the problem of ambiguity in any legal system involves establishing procedures capable of interpreting rules and their application to specific cases on the basis of legal reasoning.

The result is "an effort to gain assent to . . . judgments on reasoned rather than idiosyncratic grounds" (Kratochwil 1989, 119). International governance in this view cannot be justified legally on the basis of "bargaining" alone; it also requires a process of "arguing" against the background of commonly accepted legal norms.[11] The major instrument for establishing these argumentative procedures is the delegation of "authority to designated third parties—including courts, arbitrators and administrative organizations—to implement agreements" (Abbott et al. 2000, 415). Such delegation encompasses a number of tasks including fact finding, dispute settlement, and rule refinement in the process of rule application, with each task being restricted by the principles and terms of the agreement. The greater the autonomy of designated authorities regarding each of the three tasks, the greater the extent of juridification.

*Internalization* constitutes the second component of legalization.[12] Here, we build on the theory of the internalization of law, which asserts that norms operating above and beyond national societies can attain full legal status only when those to whom they are addressed internalize them (Koh 1997, 2645–2658; Raustiala 1995). Legal internalization refers here to the fact that rules and norms of conduct, developed outside of the jurisdiction of individual states, directly affect the behavior of their addressees. This leads to a situation in which "enforcement through domestic courts" becomes the principal means by which compliance with international rules and regulations is attained (O'Connell 1995, 5–7; Alter 2001). In its strongest form, legal internalization is based, above all, on two pillars that are best illustrated by EU practice: the supremacy of European law over the law of individual members and the binding effect of European Court of Justice case law, via the preliminary reference procedure contained in Article 234 (former Article 177) of the EU treaties. These features ensure that European law possesses an unquestionable and unquestioned validity throughout all the member states, so that community provisions become an inseparable part of the body of laws valid for all EU citizens (Weiler 1993).[13] The deeper the internalization of international rules and regulations becomes, the more likely it is that individual states will comply with them.

Summing up, legalization encompasses double processes of juridification and internalization. These components of legalization are closely related to one another other; interaction is a central feature of their dynamics (Stone, Sweet, and Caporaso 1998).

This brings us again to an examination of the empirical evidence regarding environmental regimes included in the IRD. According to the theory, the core of *juridification* consists of a delegation of the power to develop and interpret rules to bodies possessing some degree of autonomy. Courts or courtlike dispute-settlement bodies operating at the supranational or transnational level are the most important arrangements in this respect. Unfortunately, the IRD contains no direct information about dispute-settlement bodies. When we developed the IRD Data Protocol, dispute-settlement bodies did not figure prominently either in theoretical debates about international institutions or in the practice of international environmental politics. This has changed now. The theoretical debate described above is among the liveliest in the study of international institutions. In the world of international environmental policy, the last decade has witnessed the rise of more sophisticated procedures pertaining to compliance with the provisions of environmental treaties, such as the Montreal Protocol on Substances That Deplete the Ozone Layer, the protocols to the Geneva Convention on Long-Range Transboundary Air Pollution, the Basel Protocol on Liability and Compensation for Damage Resulting from Transboundary Movements of Hazardous Wastes and Their Disposal, and the Aarhus Convention on Access to Information, Public Participation in Decision-Making and Access to Justice in Environmental Matters (see Ehrmann 2002). Additional procedures of this sort are associated with the Cartagena Protocol on Biosafety, the Rotterdam Convention on the Prior Informed Consent Procedure for Certain Hazardous Chemicals and Pesticides in

Table 3.7
Regime secretariats (RA 29)

| | | ESTABLISHED SECRETARIAT | | | | | | | | | | | | | | | |
|---|---|---|---|---|---|---|---|---|---|---|---|---|---|---|---|---|---|
| | | All cases | | No entry | | 0 | | 1 | | 2 | | 3 | | 4 | | 5 | |
| CONFORMITY | 1 | 18 | 12.5% | 0 | 0.0% | 0 | 0.0% | 5 | 35.7% | 9 | 12.0% | 4 | 12.5% | 0 | 0.0% | 0 | 0.0% |
| | 2 | 72 | 50.0% | 2 | 100.0% | 1 | 20.0% | 8 | 57.2% | 39 | 52.0% | 8 | 25.0% | 0 | 0.0% | 14 | 93.3% |
| | 3 | 38 | 26.4% | 0 | 0.0% | 2 | 40.0% | 1 | 7.1% | 17 | 22.7% | 17 | 53.1% | 0 | 0.0% | 1 | 6.7% |
| | 4 | 13 | 9.0% | 0 | 0.0% | 1 | 20.0% | 0 | 0.0% | 10 | 13.3% | 2 | 6.3% | 0 | 0.0% | 0 | 0.0% |
| | 5 | 3 | 2.1% | 0 | 0.0% | 1 | 20.0% | 0 | 0.0% | 0 | 0.0% | 1 | 3.1% | 1 | 100.0% | 0 | 0.0% |
| Total | | 144 | 100.0% | 2 | 100.0% | 5 | 100.0% | 14 | 100.0% | 75 | 100.0% | 32 | 100.0% | 1 | 100.0% | 15 | 100.0% |
| | | All cases | | No entry | | 0 | | 1 | | 2 | | 3 | | 4 | | 5 | |
| CONFORMITY CAUSAL | 1 | 8 | 6% | 0 | 0.0% | 4 | 80.0% | 2 | 14.0% | 2 | 3.0% | 0 | 0.0% | 0 | 0.0% | 0 | 0.0% |
| | 2 | 39 | 27% | 2 | 100.0% | 0 | 0.0% | 0 | 0.0% | 19 | 25.0% | 8 | 25.0% | 1 | 100.0% | 9 | 60.0% |
| | 3 | 87 | 60% | 0 | 0.0% | 1 | 20.0% | 12 | 86.0% | 44 | 59.0% | 24 | 75.0% | 0 | 0.0% | 6 | 40.0% |
| | 4 | 0 | 0.0% | 0 | 0.0% | 0 | 0.0% | 0 | 0.0% | 0 | 0.0% | 0 | 0.0% | 0 | 0.0% | 0 | 0.0% |
| | 5 | 10 | 7% | 0 | 0.0% | 0 | 0.0% | 0 | 0.0% | 10 | 13.0% | 0 | 0.0% | 0 | 0.0% | 0 | 0.0% |
| Total | | 144 | 100% | 2 | 100.0% | 5 | 100.0% | 14 | 100.0% | 75 | 100.0% | 32 | 100.0% | 1 | 100.0% | 15 | 100.0% |

ESTABLISHED SECRETARIAT: 0 = not applicable, 1 = no secretariat established, 2 = regime has a secretariat of its own operating independently, 3 = intergovernmental organization performs the secretariat's functions, 4 = nongovernmental organization performs the secretariat's functions, 5 = a nation-state performs the secretariat's functions

CONFORMITY: 1 = behavior exceeds regime requirements, 2 = behavior meets regime requirements, 3 = behavior conforms with some requirements but not all, 4 = behavior conforms some (but not all) of the time and/or to some degree but not completely, 5 = behavior does not conform at all CONFORMITY_CAUSAL: 1 = little or no causal impact, 2 = modest causal influence, 3 = large causal influence, 4 = negative causal influence, 5 = don't know

International Trade, the International Treaty on Plant Genetic Resources for Food and Agriculture, and the Stockholm Convention on Persistent Organic Pollutants (see also Lefeber 2002).

Since these recent developments are not considered in the IRD, we turn to data on secretariats and decision-making bodies in order to assess the compliance effects of delegating authority to interpret and develop rules to agents possessing some degree of autonomy (table 3.7). On a general level, these data demonstrate that the presence of autonomous bodies does increase compliance.

The presence of a secretariat or similar organization with some autonomy is conducive to compliance but not a major force. The coders responded to the question: "Did the members of the regime establish a secretariat for the regime as a whole or any of its elements?" In 107 cases, the regime has a secretariat of its own (75) or an arrangement under which secretariat functions are performed by an intergovernmental organization (32).[14] On the other hand, the coders found 29 cases with no secretariat or with a member state performing the secretariat functions. As table 3.7 indicates, the contribution of secretariats or international organizations with some autonomy to compliance rates is negligible. Cases without secretariats produced high scores in terms of behavior exceeding the regime prescriptions as well.

The secretariat's level of independence also seems unimportant. Secretariats with high or strong levels of independence are not associated with better compliance rates than secretariats with little or no independence. The best compliance scores occur in cases where secretariats have *no* independence (71.4 percent had good or excessive compliance). But asked to assess the causal impact of secretariats, the coders did not identify any causal connection between the absence of a secretariat and excessive compliance. On the contrary, they found a causal connection between the existence of a secretariat and high rates of compliance. In 78.2 percent of the cases in which coders gave this variable a large causal-impact score, a secretariat was established (50.6 percent) or the relevant functions were performed by an intergovernmental organization (27.6 percent). The corresponding score for "modest causal influence" is 69.2 percent, while the coders found little or no causal impact in only 25 percent of those cases where an independent agent performed the secretariat

**Table 3.8**
Causal impact of secretariats

| | No secretariat | With secretariat function |
|---|---|---|
| No, little or negative causal influence | 2 | 2 |
| At least modest influence | 12 | 97 |

*Note:* Contingency coefficient $C = 0.21$

function.[15] Collapsing the data into only two values for each dimension (table 3.8) makes the correlation even clearer.

We conclude that successful regimes do not always require a body that can perform some regime functions autonomously. Some regimes succeed in the absence of a secretariat. But the creation of an independent secretariat does help to overcome some problems that limit compliance. Evidently, member states are able to judge whether or not an independent secretariat is needed.

The evidence relating to *autonomous decision-making bodies*—in contrast to secretariats—is stronger (table 3.9). The data in the IRD indicate clearly that the presence of a standing decision-making body rather than an ad hoc body or regular meetings of a Conference of the Parties increases compliance. The share of cases where behavior meets or exceeds requirements is 69.3 percent when such a decision-making body is present, compared to 55.2 percent when the Conference of the Parties is the only decision-making body. Again, collapsing the data into two binary variables highlights this association (table 3.10). The qualitative statements of the coders about these issues substantiate this conclusion.[16]

To some extent, we can treat the precision with which rules are formulated as a substitute for the delegation of the authority to interpret and refine rules to agents possessing a degree of autonomy. As rules become more precise, the probability of divergent interpretations resulting from ambiguity declines along with the need for dispute-settlement procedures. In fact, Abbott et al. (2000, 401) treat the precision of rules as a defining feature of legalization. The question in the IRD addressing this subject is: "Are the regime's substantive rules generally precise and easy to interpret in the sense that they call for well-defined actions, or are they

Table 3.9
Decision-making bodies (RA 33)

| | | DECISION-MAKING BODY | | | | | | | | | | | | | | | |
|---|---|---|---|---|---|---|---|---|---|---|---|---|---|---|---|---|---|
| | | All cases | | No entry | | 0 | | 1 | | 2 | | 3 | | 4 | | 5 | |
| CONFORMITY | 1 | 85 | 21.5% | 14 | 15.9% | 0 | 0.0% | 13 | 13.7% | 5 | 12.8% | 44 | 30.1% | 9 | 45.0% | 0 | 0.0% |
| | 2 | 174 | 43.9% | 48 | 54.5% | 4 | 100% | 44 | 46.3% | 12 | 30.8% | 52 | 35.6% | 10 | 50.0% | 4 | 100.0% |
| | 3 | 84 | 21.2% | 18 | 20.5% | 0 | 0.0% | 25 | 26.3% | 19 | 48.7% | 21 | 14.4% | 1 | 5.0% | 0 | 0.0% |
| | 4 | 45 | 11.4% | 5 | 5.7% | 0 | 0.0% | 12 | 12.6% | 1 | 2.6% | 27 | 18.5% | 0 | 0.0% | 0 | 0.0% |
| | 5 | 8 | 2.0% | 3 | 3.4% | 0 | 0.0% | 1 | 1.1% | 2 | 5.1% | 2 | 1.4% | 0 | 0.0% | 0 | 0.0% |
| Total | | 396 | 100.0% | 88 | 100.0% | 4 | 100.0% | 17 | 100.0% | 39 | 100.0% | 146 | 100.0% | 20 | 100.0% | 4 | 100.0% |
| | | All cases | | No entry | | 0 | | 1 | | 2 | | 3 | | 4 | | 5 | |
| CONFORMITY_ CAUSAL | 1 | 23 | 5.8% | 9 | 10.2% | 0 | 0.0% | 5 | 5.3% | 4 | 10.3% | 5 | 3.4% | 0 | 0.0% | 0 | 0.0% |
| | 2 | 84 | 34.4% | 21 | 23.9% | 4 | 100.0% | 21 | 22.1% | 5 | 12.8% | 33 | 22.6% | 0 | 0.0% | 0 | 0.0% |
| | 3 | 244 | 61.6% | 50 | 56.8% | 0 | 0.0% | 63 | 66.3% | 29 | 74.4% | 81 | 55.5% | 17 | 85.0% | 4 | 100.0% |
| | 4 | 0 | 0.0% | 0 | 0.0% | 0 | 0.0% | 0 | 0.0% | 0 | 0.0% | 0 | 0.0% | 0 | 0.0% | 0 | 0.0% |
| | 5 | 45 | 11.4% | 8 | 9.1% | 0 | 0.0% | 6 | 6.3% | 1 | 2.6% | 27 | 18.5% | 3 | 15.0% | 0 | 0.0% |
| Total | | 396 | 100% | 88 | 100.0% | 4 | 100.0% | 95 | 100.0% | 39 | 100.0% | 146 | 100.0% | 20 | 100.0% | 4 | 100.0% |

DECISION-MAKING BODY: 0 = not applicable, 1 = regular meeting of conference of parties, 2 = ad hoc meetings of conference of parties, 3 = standing subsidiary body, 4 = ad hoc subsidiary body, 5 = don't know
CONFORMITY: 1 = behavior exceeds regime requirements, 2 = behavior meets regime requirements, 3 = behavior conforms with some requirements but not all, 4 = behavior conforms some (but not all) of the time and/or to some degree but not completely, 5 = behavior does not conform at all
CONFORMITY_CAUSAL: 1 = little or no causal impact, 2 = modest causal influence, 3 = large causal influence, 4 = negative causal influence, 5 = don't know

**Table 3.10**
Autonomy and compliance rates

| | Conference of the Parties | Subsidiary Body |
|---|---|---|
| Bad compliance 3–5 | 60 | 49 |
| Good compliance | 74 | 115 |

*Note:* Contingency coefficient C = 0.15

ambiguous and indeterminate?" (RA 12). The scale associated with this variable ranges from 1 to 5, with 1 being the most precise.

The association between the precision of rules and compliance rates is positive but not strong (table 3.11). The overall effect is attributable mainly to the existence of rules that are very precise and easy to interpret. Precise rules (i.e., those that score 1) produce good or excessive compliance in 62.6 percent of the cases, compared to the overall average of 53.3 percent. But note that there is no positive effect for rules that score 2 on the scale (48.8 percent), and there is a distinct negative effect for rules scoring 3 (30.7 percent). What is more, although precise rules increase compliance, they do not enhance behavior that exceeds compliance. Rule precision reduces deviations from the prescribed behavior in both directions (i.e., overcompliance as well as noncompliance). The judgments of the coders regarding causality confirm these conclusions regarding the relationship between the precision of rules and compliance rates.

Overall, this assessment is compatible with the hypothesis that juridification improves compliance rates. Both the delegation of authority to interpret and refine rules to agents with some degree of autonomy and the formulation of precise rules are conducive to compliance. But the evidence from the IRD is not particularly strong regarding these links.

The IRD does not contain direct measures of internalization, the second component of legalization. So far, international environmental regimes have not empowered individuals to direct complaints against their own states. The only question in the IRD that we can construe as an indirect indicator of legal internalization is RA 11, which asks: "Are the regime's substantive rules legally binding on the members, or do they have the character of soft law (e.g. ministerial declarations, codes of conduct)?"

Interestingly, this variable seems to have no effect on compliance rates at all (table 3.12). Levels of compliance mirror the general distribution between legally binding and soft-law rules—about 80 to 20 percent—with remarkably little variation.

To sum up, although the evidence in the IRD is not ideal for testing this theoretical perspective, reducing the ambiguity of rules through legalization does raise compliance rates. For international environmental regimes, at least, juridification is more effective than legal internalization. Both the presence of implementing agencies with some autonomy and the precision of the rules improve compliance.

### 3.3.3 Legitimacy against Nonacceptance

This perspective treats reservations about the normative validity of a rule as the most significant source of noncompliance, a view that points to the search for legitimacy as the key to solving compliance problems. The assumption here is that general precepts regarding justice and fairness must play a recognizable part in making and applying rules in order for them to be accepted as legitimate. The *manner* in which norms are generated and applied distinguishes legitimate rules from those that lack legitimacy. To be legitimate, rules and regulations should emerge from legitimate norm-forming processes *and* be applied in a way that demonstrates a rational linkage to their goals and to certain general principles of fairness or justice (Dworkin 1986; Habermas 1994). This perspective highlights the importance of a clear link between legitimacy and compliance, even more than the legalization perspective does.

On this account, the discourses used to justify rules must conform to the principles of rational discourse and apply to all the addressees; those who apply the rules must embrace these discourses. Thus, the legitimacy of a rule is a function of the extent to which decision making regarding the rule is judged to be fair. Subjects are likely to regard a rule as fair when they have an opportunity to participate in decision making relating to the rule and when the rule is not systematically biased in favor of certain interests or interest groups. Although participation and impartiality

**Table 3.11**
Precision (RA 12)

| | | RULE GENERALLY PRECISE AND EASY TO INTERPRET (RULE_PRECISE) | | | | | | | | | | | | | | | |
|---|---|---|---|---|---|---|---|---|---|---|---|---|---|---|---|---|---|
| | | All cases | | No entry | | 0 | | 1 | | 2 | | 3 | | 4 | | 5 | |
| CONFORMITY | 1 | 121 | 12.6% | 10 | 41.7% | 0 | 0.0% | 42 | 11.1% | 38 | 14.2% | 11 | 7.2% | 12 | 13.5% | 8 | 24.2% |
| | 2 | 391 | 40.7% | 14 | 58.3% | 6 | 42.9% | 196 | 51.6% | 92 | 34.5% | 36 | 23.5% | 37 | 41.6% | 10 | 30.3% |
| | 3 | 332 | 34.6% | 0 | 0.0% | 6 | 42.9% | 110 | 28.9% | 107 | 40.1% | 62 | 40.5% | 35 | 39.3% | 12 | 36.4% |
| | 4 | 86 | 9.0% | 0 | 0.0% | 1 | 7.1% | 28 | 7.4% | 24 | 9.0% | 28 | 18.3% | 3 | 3.4% | 2 | 6.1% |
| | 5 | 30 | 3.1% | 0 | 0.0% | 1 | 7.1% | 4 | 1.1% | 6 | 2.2% | 16 | 10.5% | 2 | 2.2% | 1 | 3.0% |
| Total | | 960 | 100.0% | 24 | 100.0% | 14 | 100.0% | 380 | 100.0% | 267 | 100.0% | 153 | 100.0% | 89 | 100.0% | 33 | 100.0% |
| | | | | | | | | | | | | | | | | | |
| | | All cases | | No entry | | 0 | | 1 | | 2 | | 3 | | 4 | | 5 | |
| CONFORMITY_CAUSAL | 1 | 44 | 4.6% | 0 | 0.0% | 4 | 28.6% | 3 | 0.8% | 11 | 4.1% | 7 | 4.6% | 10 | 11.2% | 9 | 27.3% |
| | 2 | 234 | 24.4% | 0 | 0.0% | 4 | 28.6% | 96 | 25.3% | 32 | 12.0% | 56 | 36.6% | 34 | 38.2% | 12 | 36.4% |
| | 3 | 642 | 66.9% | 19 | 79.2% | 4 | 28.6% | 272 | 71.6% | 213 | 79.8% | 79 | 51.6% | 44 | 49.4% | 11 | 33.3% |
| | 4 | 0 | 0.0% | 0 | 0.0% | 0 | 0.0% | 0 | 0.0% | 0 | 0.0% | 0 | 0.0% | 0 | 0.0% | 0 | 0.0% |
| | 5 | 40 | 4.2% | 5 | 20.8% | 2 | 14.3% | 9 | 2.4% | 11 | 4.1% | 11 | 7.2% | 1 | 1.1% | 1 | 3.0% |
| Total | | 960 | 100.0% | 24 | 100.0% | 14 | 100.0% | 380 | 100.0% | 267 | 100.0% | 153 | 100.0% | 89 | 100.0% | 33 | 100.0% |

RULE_PRECISE: 0 = not applicable, 1 = precise and easy to interpret, 2 = between 1 and 3 on the scale, 3 = medium, 4 = between 3 and 5 on the scale, 5 = ambiguous and indeterminate

CONFORMITY: 1 = behavior exceeds regime requirements, 2 = behavior meets regime requirements, 3 = behavior conforms with some requirements but not all, 4 = behavior conforms some (but not all) of the time and/or to some degree but not completely, 5 = behavior does not conform at all

CONFORMITY_CAUSAL: 1 = little or no causal impact, 2 = modest causal influence, 3 = large causal influence, 4 = negative causal influence, 5 = don't know

**Table 3.12**
Legally binding rules or soft law (RA 11)

| | | RULES LEGALLY BINDING OR SOFT LAW (RULE_BINDING) | | | | | | | | | | | |
|---|---|---|---|---|---|---|---|---|---|---|---|---|---|
| | | All cases | | No entry | | 0 | | 1 | | 2 | | 3 | |
| CONFORMITY | 1 | 124 | 12.3% | 0 | 0.0% | 4 | 26.7% | 104 | 13.5% | 16 | 8.2% | 0 | 0.0% |
| | 2 | 433 | 43.0% | 1 | 50.0% | 7 | 46.7% | 310 | 40.2% | 92 | 47.2% | 23 | 100.0% |
| | 3 | 335 | 33.3% | 0 | 0.0% | 2 | 13.3% | 258 | 33.4% | 75 | 38.5% | 0 | 0.0% |
| | 4 | 86 | 8.5% | 1 | 50.0% | 1 | 6.7% | 77 | 10.0% | 7 | 3.6% | 0 | 0.0% |
| | 5 | 29 | 2.9% | 0 | 0.0% | 1 | 6.7% | 23 | 3.0% | 5 | 2.6% | 0 | 0.0% |
| Total | | 1007 | 100.0% | 2 | 100.0% | 15 | 100.0% | 772 | 100.0% | 195 | 100.0% | 23 | 100.0% |
| | | All cases | | No entry | | 0 | | 1 | | 2 | | 3 | |
| CONFORMITY_ CAUSAL | 1 | 44 | 4.4% | 0 | 0.0% | 2 | 13.3% | 35 | 4.5% | 7 | 3.6% | 0 | 0.0% |
| | 2 | 246 | 24.4% | 1 | 50.0% | 3 | 20.0% | 192 | 24.9% | 50 | 25.6% | 0 | 0.0% |
| | 3 | 677 | 67.2% | 0 | 0.0% | 8 | 53.3% | 524 | 67.9% | 122 | 62.6% | 23 | 100.0% |
| | 4 | 0 | 0.0% | 0 | 0.0% | 0 | 0.0% | 0 | 0.0% | 0 | 0.0% | 0 | 0.0% |
| | 5 | 40 | 4.0% | 1 | 50.0% | 2 | 13.3% | 21 | 2.7% | 16 | 8.2% | 0 | 0.0% |
| Total | | 1007 | 100.0% | 2 | 100.0% | 15 | 100.0% | 772 | 100.0% | 195 | 100.0% | 23 | 100.0% |

RULE_BINDING: 0 = not applicable, 1 = rule is legally binding, 2 = rule is soft law/not legally binding, 3 = don't know
CONFORMITY: 1 = behavior exceeds regime requirements, 2 = behavior meets regime requirements, 3 = behavior conforms with some requirements but not all, 4 = behavior conforms some (but not all) of the time and/or to some degree but not completely, 5 = behavior does not conform at all know
CONFORMITY_CAUSAL: 1 = little or no causal impact, 2 = modest causal influence, 3 = large causal influence, 4 = negative causal influence, 5 = don't know

are not the only ingredients of a procedural theory of justice, they enjoy broad popular support (Tyler 1990).

At a minimum, participation and impartiality must apply to all those who are immediate *addressees* of a rule or regulation. Ideally, the final *targets* of rules should be able to participate directly or indirectly in making and applying rules, and all others who are *affected* should partake of the relevant discourses. Public participation reduces the danger that social power will be transformed into administrative power. Thus, the legitimacy perspective emphasizes two determinants of compliance: the inclusion of all subjects of rules and regulations in the decision-making process and the involvement of all associated regulatory targets and those affected by rules and regulations in the decision-making process.

With regard to the inclusion of relevant actors, the most pertinent question in the IRD (RF 46) asks: "Were there states not participating in the negotiation process that other actors believed should have participated?" (See table 3.13a.)

The absence of important states in the decision-making process does affect compliance rates negatively (table 3.13b). When no important states are missing, the likelihood of good compliance or behavior exceeding compliance increases to 66.7 percent from an average of 60.8 percent. Equally important is the observation that it is more important in these terms to include all those states treated as contributors to the problem than to include those deemed important in terms of problem solving.

At the same time, asymmetries in power, which may indicate the existence of a highly uneven pattern of decision making, have no significant impact on compliance rates. Approximately even distributions correlate only weakly with higher rates of compliance; the coders generally judged this factor to be irrelevant to compliance (table 3.14). Nor did different forms of domination occurring during regime formation affect compliance.

The IRD does not differentiate explicitly among subjects, targets, and affected parties with regard to decision-making processes. We therefore sought an indirect measure of affected parties. Specifically, we considered the IRD question (RD 49) that asks: "What roles did non-state actors play in the negotiations?"

**Table 3.13a**
Missing states (RF 46)

<table>
<tr><td colspan="2"></td><td colspan="10">STATES NOT PATICIPATING IN NEGOTIATION PROCESS THAT OTHER ACTORS BELIEVED SHOULD HAVE PARTICIPATED (NEGOTIATE_NOT_PARTICIPATE)</td></tr>
<tr><td colspan="2"></td><td colspan="2">All cases</td><td colspan="2">0 = Not applicable</td><td colspan="2">1 = No</td><td colspan="2">2 = Yes</td><td colspan="2">3 = Don't know</td></tr>
<tr><td rowspan="5">CONFORMITY</td><td>1</td><td>15</td><td>12.0%</td><td>0</td><td>0.0%</td><td>12</td><td>15.4%</td><td>3</td><td>12.0%</td><td>0</td><td>0.0%</td></tr>
<tr><td>2</td><td>61</td><td>48.8%</td><td>7</td><td>100.0%</td><td>40</td><td>51.3%</td><td>12</td><td>48.0%</td><td>2</td><td>13.3%</td></tr>
<tr><td>3</td><td>33</td><td>26.4%</td><td>0</td><td>0.0%</td><td>18</td><td>23.1%</td><td>6</td><td>24.0%</td><td>9</td><td>60.0%</td></tr>
<tr><td>4</td><td>13</td><td>10.4%</td><td>0</td><td>0.0%</td><td>7</td><td>9.0%</td><td>3</td><td>12.0%</td><td>3</td><td>20.0%</td></tr>
<tr><td>5</td><td>3</td><td>2.4%</td><td>0</td><td>0.0%</td><td>1</td><td>1.3%</td><td>1</td><td>4.0%</td><td>1</td><td>6.7%</td></tr>
<tr><td colspan="2">Total</td><td>125</td><td>100.0%</td><td>7</td><td>100.0%</td><td>78</td><td>100.0%</td><td>25</td><td>100.0%</td><td>15</td><td>100.0%</td></tr>
<tr><td colspan="12"></td></tr>
<tr><td colspan="2"></td><td colspan="2">All cases</td><td colspan="2">0 = Not applicable</td><td colspan="2">1 = No</td><td colspan="2">2 = Yes</td><td colspan="2">3 = Don't know</td></tr>
<tr><td rowspan="5">CONFORMITY_ CAUSAL</td><td>1</td><td>8</td><td>6.40%</td><td>0</td><td>0.0%</td><td>4</td><td>5.1%</td><td>2</td><td>8.0%</td><td>2</td><td>13.3%</td></tr>
<tr><td>2</td><td>31</td><td>24.80%</td><td>2</td><td>28.6%</td><td>20</td><td>25.6%</td><td>7</td><td>28.0%</td><td>2</td><td>13.3%</td></tr>
<tr><td>3</td><td>77</td><td>61.60%</td><td>5</td><td>71.4%</td><td>48</td><td>61.5%</td><td>16</td><td>64.0%</td><td>8</td><td>53.3%</td></tr>
<tr><td>4</td><td>0</td><td>0.00%</td><td>0</td><td>0.0%</td><td>0</td><td>0.0%</td><td>0</td><td>0.0%</td><td>0</td><td>0.0%</td></tr>
<tr><td>5</td><td>9</td><td>7.20%</td><td>0</td><td>0.0%</td><td>6</td><td>7.7%</td><td>0</td><td>0.0%</td><td>3</td><td>20.0%</td></tr>
<tr><td colspan="2">Total</td><td>125</td><td>100.00%</td><td>7</td><td>100.0%</td><td>78</td><td>100.0%</td><td>25</td><td>100.0%</td><td>15</td><td>100.0%</td></tr>
</table>

NEGOTIATE_NOT_PARTICIPATE: 0 = not applicable, 1 = no, 2 = yes, 3 = don't know
CONFORMITY: 1 = behavior exceeds regime requirements, 2 = behavior meets regime requirements, 3 = behavior conforms with some requirements but not all, 4 = behavior conforms some (but not all) of the time and/or to some degree but not completely, 5 = behavior does not conform at all
CONFORMITY_CAUSAL: 1 = little or no causal impact, 2 = modest causal influence, 3 = large causal influence, 4 = negative causal influence, 5 = don't know

**Table 3.13b**
Missing states: Problem makers and problem solvers (RF 46)

| | | STATES IMPORTANT AS CONTRIBUTORS OR IMPORTANT IN TERMS OF PROBLEM SOLVING (NEGOTIATE_NOT_STATE_TYPE) | | | | | | | | | | | | | |
|---|---|---|---|---|---|---|---|---|---|---|---|---|---|---|---|
| | | All cases | | No entry | | 0 | | 1 | | 2 | | 3 | | 4 | |
| CONFORMITY | 1 | 7 | 12.5% | 0 | 0.0% | 3 | 50.0% | 0 | 0.0% | 4 | 66.7% | 0 | 0.0% | 0 | 0.0% |
| | 2 | 27 | 48.2% | 2 | 50.0% | 3 | 50.0% | 14 | 48.3% | 2 | 33.3% | 6 | 75.0% | 0 | 0.0% |
| | 3 | 15 | 26.8% | 1 | 25.0% | 0 | 0.0% | 12 | 41.4% | 0 | 0.0% | 1 | 12.5% | 1 | 33.3% |
| | 4 | 5 | 8.9% | 0 | 0.0% | 0 | 0.0% | 3 | 10.3% | 0 | 0.0% | 1 | 12.5% | 1 | 33.3% |
| | 5 | 2 | 3.6% | 1 | 25.0% | 0 | 0.0% | 0 | 0.0% | 0 | 0.0% | 0 | 0.0% | 1 | 33.3% |
| Total | | 56 | 100.0% | 4 | 100.0% | 6 | 100.0% | 29 | 100.0% | 6 | 100.0% | 8 | 100.0% | 3 | 100.0% |
| | | All cases | | No entry | | 0 | | 1 | | 2 | | 3 | | 4 | |
| CONFORMITY_CAUSAL | 1 | 4 | 7.10% | 2 | 50.0% | 0 | 0.0% | 2 | 6.9% | 0 | 0.0% | 0 | 0.0% | 0 | 0.0% |
| | 2 | 11 | 19.60% | 1 | 25.0% | 1 | 16.7% | 3 | 10.3% | 2 | 33.3% | 3 | 37.5% | 1 | 33.3% |
| | 3 | 41 | 73.20% | 1 | 25.0% | 5 | 83.3% | 24 | 82.8% | 4 | 66.7% | 5 | 62.5% | 2 | 66.7% |
| | 4 | 0 | 0.00% | 0 | 0.0% | 0 | 0.0% | 0 | 0.0% | 0 | 0.0% | 0 | 0.0% | 0 | 0.0% |
| | 5 | 0 | 0.00% | 0 | 0.0% | 0 | 0.0% | 0 | 0.0% | 0 | 0.0% | 0 | 0.0% | 0 | 0.0% |
| Total | | 56 | 100.0% | 4 | 100.0% | 6 | 100.0% | 29 | 100.0% | 6 | 100.0% | 8 | 100.0% | 3 | 100.0% |

NEGOTIATE_NOT_STATE_TYPE: 0 = not applicable, 1 = important contributor to problem, 2 = important in problem solving, 3 = important contributor to problem and important in terms of problem solving, 4 = don't know

CONFORMITY: 1 = behavior exceeds regime requirements, 2 = behavior meets regime requirements, 3 = behavior conforms with some requirements but not all, 4 = behavior conforms some (but not all) of the time and/or to some degree but not completely, 5 = behavior does not conform at all

CONFORMITY_CAUSAL: 1 = little or no causal impact, 2 = modest causal influence, 3 = large causal influence, 4 = negative causal influence, 5 = don't know

**Table 3.14**
Issue-area specific power (RF 19)

| | | SYMMETRICAL VS. ASYMMETRICAL DISTRIBUTIONS OF POWER (POWER_SETTING_SYMMETRY) | | | | | | | | | | | | | | | |
|---|---|---|---|---|---|---|---|---|---|---|---|---|---|---|---|---|---|
| | | All cases | | No entry | | 0 | | 1 | | 2 | | 3 | | 4 | | 5 | |
| CONFORMITY | 1 | 18 | 13.8% | 0 | 0.0% | 2 | 25.0% | 11 | 26.8% | 4 | 8.9% | 1 | 3.2% | 0 | 0.0% | 0 | 0.0% |
| | 2 | 62 | 47.7% | 2 | 100.0% | 4 | 50.0% | 15 | 36.6% | 22 | 48.9% | 17 | 54.8% | 0 | 0.0% | 2 | 100.0% |
| | 3 | 34 | 26.2% | 0 | 0.0% | 2 | 25.0% | 11 | 26.8% | 15 | 33.3% | 6 | 19.4% | 0 | 0.0% | 0 | 0.0% |
| | 4 | 13 | 10.0% | 0 | 0.0% | 0 | 0.0% | 3 | 7.3% | 3 | 6.7% | 6 | 19.4% | 1 | 100.0% | 0 | 0.0% |
| | 5 | 3 | 2.3% | 0 | 0.0% | 0 | 0.0% | 1 | 2.4% | 1 | 2.2% | 1 | 3.2% | 0 | 0.0% | 0 | 0.0% |
| **Total** | | **130** | **100.0%** | **2** | **100.0%** | **8** | **100.0%** | **41** | **100.0%** | **45** | **100.0%** | **31** | **100.0%** | **1** | **100.0%** | **2** | **100.0%** |
| | | All cases | | No entry | | 0 | | 1 | | 2 | | 3 | | 4 | | 5 | |
| CONFORMITY_ CAUSAL | 1 | 8 | 6.2% | 0 | 0.0% | 0 | 0.0% | 6 | 14.6% | 1 | 2.2% | 1 | 3.2% | 0 | 0.0% | 0 | 0.0% |
| | 2 | 32 | 24.6% | 0 | 0.0% | 2 | 25.0% | 5 | 12.2% | 12 | 26.7% | 13 | 41.9% | 0 | 0.0% | 0 | 0.0% |
| | 3 | 80 | 61.5% | 1 | 50.0% | 3 | 37.5% | 27 | 65.9% | 30 | 66.7% | 16 | 51.6% | 1 | 100.0% | 2 | 100.0% |
| | 4 | 0 | 0.0% | 0 | 0.0% | 0 | 0.0% | 0 | 0.0% | 0 | 0.0% | 0 | 0.0% | 0 | 0.0% | 0 | 0.0% |
| | 5 | 10 | 7.7% | 1 | 50.0% | 3 | 37.5% | 3 | 7.3% | 2 | 4.4% | 1 | 3.2% | 0 | 0.0% | 0 | 0.0% |
| **Total** | | **130** | **100.0%** | **2** | **100.0%** | **8** | **100.0%** | **41** | **100.0%** | **45** | **100.0%** | **31** | **100.0%** | **1** | **100.0%** | **2** | **100.0%** |

POWER_SETTING_SYMMETRY: 0 = not applicable, 1 = completely even distribution, 2 = slightly uneven distribution, 3 = considerable unevenness, 4 = highly uneven distribution, 5 = issue-specific hegemon present
CONFORMITY: 1 = behavior exceeds regime requirements, 2 = behavior meets regime requirements, 3 = behavior conforms with some requirements but not all, 4 = behavior conforms some (but not all) of the time and/or to some degree but not completely, 5 = behavior does not conform at all
CONFORMITY_CAUSAL: 1 = little or no causal impact, 2 = modest causal influence, 3 = large causal influence, 4 = negative causal influence, 5 = don't know

**Table 3.15**
Role of nonstate actors (RF 49)

| | | ROLE OF NON-STATE ACTORS IN NEGOTIATIONS (NEGOTIATE_NON_STATE_ROLE) | | | | | | | | | | | | | | | |
|---|---|---|---|---|---|---|---|---|---|---|---|---|---|---|---|---|---|
| | | All cases | | No entry | | 0 | | 1 | | 2 | | 3 | | 4 | | 5 | |
| CONFORMITY | 1 | 129 | 12.0% | 1 | 0.9% | 29 | 10.7% | 7 | 10.3% | 0 | 0.0% | 44 | 17.6% | 48 | 15.9% | 0 | 0.0% |
| | 2 | 503 | 46.9% | 56 | 50.5% | 116 | 42.6% | 45 | 66.2% | 20 | 62.5% | 99 | 39.6% | 144 | 47.8% | 23 | 60.5% |
| | 3 | 292 | 27.2% | 35 | 31.5% | 80 | 29.4% | 9 | 13.2% | 11 | 34.4% | 77 | 30.8% | 68 | 22.6% | 12 | 31.6% |
| | 4 | 110 | 10.3% | 18 | 16.2% | 32 | 11.8% | 5 | 7.4% | 1 | 3.1% | 21 | 8.4% | 30 | 10.0% | 3 | 7.9% |
| | 5 | 38 | 3.5% | 1 | 0.9% | 15 | 5.5% | 2 | 2.9% | 0 | 0.0% | 9 | 3.6% | 11 | 3.7% | 0 | 0.0% |
| **Total** | | **1072** | **100.0%** | **111** | **100.0%** | **272** | **100.0%** | **68** | **100.0%** | **32** | **100.0%** | **250** | **100.0%** | **301** | **100.0%** | **38** | **100.0%** |

| | | All cases | | No entry | | 0 | | 1 | | 2 | | 3 | | 4 | | 5 | |
|---|---|---|---|---|---|---|---|---|---|---|---|---|---|---|---|---|---|
| CONFORMITY_CAUSAL | 1 | 48 | 4.5% | 3 | 2.7% | 16 | 5.9% | 2 | 2.9% | 3 | 9.4% | 11 | 4.4% | 11 | 3.7% | 2 | 5.3% |
| | 2 | 296 | 27.6% | 49 | 44.1% | 79 | 29.0% | 12 | 17.7% | 3 | 9.4% | 58 | 23.2% | 73 | 24.3% | 22 | 57.9% |
| | 3 | 656 | 61.2% | 49 | 44.1% | 157 | 57.7% | 51 | 75.0% | 26 | 81.3% | 169 | 67.6% | 190 | 63.1% | 14 | 36.8% |
| | 4 | 0 | 0.0% | 0 | 0.0% | 0 | 0.0% | 0 | 0.0% | 0 | 0.0% | 0 | 0.0% | 0 | 0.0% | 0 | 0.0% |
| | 5 | 72 | 6.7% | 10 | 9.0% | 20 | 7.4% | 3 | 4.4% | 0 | 0.0% | 12 | 4.8% | 27 | 9.0% | 0 | 0.0% |
| **Total** | | **1072** | **100.0%** | **111** | **100.0%** | **272** | **100.0%** | **68** | **100.0%** | **32** | **100.0%** | **250** | **100.0%** | **301** | **100.0%** | **38** | **100.0%** |

NEGOTIATE_NON_STATE_ROLE: 0 = not applicable, 1 = observer role, 2 = member of national delegation, 3 = member of negotiation body, 4 = exerted pressure inside the negotiation, 5 = exerted pressure outside the negotiation
CONFORMITY: 1 = behavior exceeds regime requirements, 2 = behavior meets regime requirements, 3 = behavior conforms with some requirements but not all, 4 = behavior conforms some (but not all) of the time and/or to some degree but not completely, 5 = behavior does not conform at all
CONFORMITY_CAUSAL: 1 = little or no causal impact, 2 = modest causal influence, 3 = large causal influence, 4 = negative causal influence, 5 = don't know

The responses of coders to this question suggest that the inclusion of NGOs as members of national delegations or as members of negotiating bodies (i.e., not just as observers) has a positive impact on compliance rates (table 3.15). Behavior meets or exceeds requirements in 53.3 percent of the cases when NGOs are observers. But this number rises to 73 percent when NGOs participate directly in the negotiation process. The data suggest that it makes no difference whether NGOs exert pressure inside or outside the formal negotiations.

From these findings, we can conclude that the absence of important states does reduce compliance and that the participation of NGOs (which we treat as a surrogate for participation by targets of regulations and those affected by regulations) increases compliance rates. Both these conclusions are compatible with hypotheses derived from the legitimacy perspective. Power asymmetries and patterns of domination, on the other hand, do not affect compliance rates negatively.

### 3.3.4 Responsiveness against Unintentional Noncompliance

This perspective treats problems of implementation as a major threat to compliance and proposes a solution featuring regulatory deliberations among experts.[17] Thus, effective rule making and rule application emerge as elements of a permanent process of interactive adjudication (Joerges 2000; Selznick 1985).

This line of analysis challenges perspectives emphasizing coercion by highlighting two empirical observations. At the international level, "Sanctioning authority is rarely granted by treaty, rarely used when granted, and likely to be ineffective when used" (Chayes and Chayes 1995, 32). What is more, proponents of this view even challenge the proposition that outside observers can easily identify cases of noncompliance. They regard regulation and compliance as a continuous process in which rules are applied and modified through a process of permanent adaptation designed to meet the requirements of effective regulation. The formation, implementation, and modification of rules constitute a continuous process that does not take the form of a linear temporal sequence.[18]

This is akin to the view that foreign-policy practitioners operate on the assumption of a general propensity of states to comply with international obligations (Chayes and Chayes 1995, 3). Proponents of this view draw particular attention to the fact that noncompliance is often unintentional. Any of a number of factors can lead to unintentional noncompliance. Ambiguity and the indeterminacy of rules is one source of unintentional noncompliance. But equally if not more important are the inability of member states to fulfill their obligations and the failure of treaties to adapt to changing conditions. Under the circumstances, this perspective expects compliance management to be successful when the parties have sufficient resources to implement rules and regulations and when the compliance-management system displays sufficient flexibility to adapt or adjust to new problems.

There are two basic ways to ensure that all regime members have sufficient capacity to implement rules. One approach is to draft rules in a manner that takes into account differences in capacity. In this vein, the IRD asks (RA 13): "Does the regime have substantive rules that differentiate among its members in terms of requirements, prohibitions, or permissions?"

The evidence in this case runs counter to our expectation. When rules differentiate among members, behavior that is fully compliant or exceeds compliance targets declines (table 3.16). While subjects comply or exceed compliance with undifferentiated rules in 54 percent of the cases, the comparable figure for cases involving differentiated rules is only 41 percent.

The other way to maximize the implementation of rules is to create mechanisms designed to build capacity for actors that have specific needs. Although 16 percent of the cases in the IRD make use of this mechanism, the effects on compliance differ from those that the responsiveness perspective projects (table 3.17). Compliance mechanisms featuring some sort of negative sanctions produce compliance or behavior exceeding compliance in over 74 percent of the cases. But performance in these terms declines to 46.7 percent in cases relying on capacity building as a compliance mechanism. The results of granting some subjects a transition period with regard to compliance are even worse. Behavior meets or exceeds compliance targets in only 21.4 percent of these cases.

These findings are surprising. They run counter not only to the responsiveness perspective regarding international regimes but also to more

**Table 3.16**
Rule differentiation (RA 13)

| | | REGIME RULES THAT DIFFERENTIATE AMONG ITS MEMBERS (RULE_DIFFERENTIATE) | | | | | | | | | | | |
|---|---|---|---|---|---|---|---|---|---|---|---|---|---|
| | | All cases | | No entry | | 0 | | 1 | | 2 | | 3 | |
| CONFORMITY | 1 | 121 | 12.7% | 10 | 29.4% | 0 | 0.0% | 105 | 13.0% | 6 | 7.1% | 0 | 0.0% |
| | 2 | 387 | 40.7% | 24 | 70.6% | 16 | 69.6% | 318 | 39.3% | 28 | 32.9% | 1 | 100.0% |
| | 3 | 331 | 34.8% | 0 | 0.0% | 1 | 4.3% | 297 | 36.7% | 33 | 38.8% | 0 | 0.0% |
| | 4 | 83 | 8.7% | 0 | 0.0% | 5 | 21.7% | 63 | 7.8% | 15 | 17.6% | 0 | 0.0% |
| | 5 | 30 | 3.2% | 0 | 0.0% | 1 | 4.3% | 26 | 3.2% | 3 | 3.5% | 0 | 0.0% |
| Total | | 952 | 100.0% | 34 | 100.0% | 23 | 100.0% | 809 | 100.0% | 85 | 100.0% | 1 | 100.0% |
| | | All cases | | No entry | | 0 | | 1 | | 2 | | 3 | |
| CONFORMITY + CAUSAL | 1 | 44 | 4.6% | 0 | 0.0% | 2 | 8.7% | 37 | 4.6% | 5 | 5.9% | 0 | 0.0% |
| | 2 | 232 | 24.4% | 1 | 2.9% | 3 | 13.0% | 192 | 23.7% | 36 | 42.4% | 0 | 0.0% |
| | 3 | 636 | 66.8% | 28 | 82.4% | 16 | 69.6% | 554 | 68.5% | 37 | 43.5% | 1 | 100.0% |
| | 4 | 0 | 0.0% | 0 | 0.0% | 0 | 0.0% | 0 | 0.0% | 0 | 0.0% | 0 | 0.0% |
| | 5 | 40 | 4.2% | 5 | 14.7% | 2 | 8.7% | 26 | 3.2% | 7 | 8.2% | 0 | 0.0% |
| Total | | 952 | 100.0% | 34 | 100.0% | 23 | 100.0% | 809 | 100.0% | 85 | 100.0% | 1 | 100.0% |

RULE_DIFFERENTIATE: 0 = not applicable, 1 = rule does not differentiate among members, 2 = rule differentiates among members, 3 = don't know

CONFORMITY: 1 = behavior exceeds regime requirements, 2 = behavior meets regime requirements, 3 = behavior conforms with some requirements but not all, 4 = behavior conforms some (but not all) of the time and/or to some degree but not completely, 5 = behavior does not conform at all

CONFORMITY_CAUSAL: 1 = little or no causal impact, 2 = modest causal influence, 3 = large causal influence, 4 = negative causal influence, 5 = don't know

**Table 3.17**
Capacity building (RA 47)

| | | COMPLIANCE MECHANISMS SOFT LAW | | | | | | | | | | | |
|---|---|---|---|---|---|---|---|---|---|---|---|---|---|
| | | All cases | | No entry | | Sum of 0–6, 10–11 | | 7 | | 8 | | 9 | |
| CONFORMITY | 1 | 32 | 17.2% | 9 | 60.0% | 18 | 14.2% | 5 | 16.7% | 0 | 0.0% | 0 | 0.0% |
| | 2 | 90 | 48.4% | 3 | 20.0% | 75 | 59.1% | 9 | 30.0% | 3 | 21.4% | 0 | 0.0% |
| | 3 | 45 | 24.2% | 1 | 6.7% | 22 | 17.3% | 15 | 50.0% | 7 | 50.0% | 0 | 0.0% |
| | 4 | 16 | 8.6% | 2 | 13.3% | 9 | 7.1% | 1 | 3.3% | 4 | 28.6% | 0 | 0.0% |
| | 5 | 3 | 1.6% | 0 | 0.0% | 3 | 2.4% | 0 | 0.0% | 0 | 0.0% | 0 | 0.0% |
| **Total** | | **186** | **100.0%** | **15** | **100.0%** | **127** | **100.0%** | **30** | **100.0%** | **14** | **100.0%** | **—** | **—** |

| | | All cases | | No entry | | Sum of 0–6, 10–11 | | 7 | | 8 | | 9 | |
|---|---|---|---|---|---|---|---|---|---|---|---|---|---|
| CONFORMITY_ CAUSAL | 1 | 9 | 4.8% | 0 | 0.0% | 8 | 6.3% | 1 | 3.3% | 0 | 0.0% | 0 | 0.0% |
| | 2 | 47 | 25.3% | 1 | 6.7% | 39 | 30.7% | 4 | 13.3% | 3 | 21.4% | 0 | 0.0% |
| | 3 | 120 | 64.5% | 14 | 6.7% | 73 | 57.5% | 25 | 83.4% | 8 | 57.1% | 0 | 0.0% |
| | 4 | 0 | 0.0% | 0 | 0.0% | 0 | 0.0% | 0 | 0.0% | 0 | 0.0% | 0 | 0.0% |
| | 5 | 10 | 5.4% | 0 | 0.0% | 7 | 5.5% | 0 | 0.0% | 3 | 21.4% | 0 | 0.0% |
| **Total** | | **186** | **100.0%** | **15** | **100.0%** | **127** | **100.0%** | **30** | **100.0%** | **14** | **100.0%** | **—** | **—** |

COMPLIANCE MECHANISMS SOFT LAW: 0 = not applicable, 1 = no compliance mechanisms, 2 = issuance of notice of violation, 3 = suspension of membership rights, 4 = exclusion from membership, 5 = imposition of military punishments, 6 = imposition of financial/economic punishments, 7 = support for capacity building to enhance compliance, 8 = granting of a transition period to active compliance, 9 = dissolution of linkages, 10 = additional compliance mechanisms, 11 = don't know

CONFORMITY: 1 = behavior exceeds regime requirements, 2 = behavior meets regime requirements, 3 = behavior conforms with some requirements but not all, 4 = behavior conforms some (but not all) of the time and/or to some degree but not completely, 5 = behavior does not conform at all

CONFORMITY_CAUSAL: 1 = little or no causal impact, 2 = modest causal influence, 3 = large causal influence, 4 = negative causal influence, 5 = don't know

general findings in the social sciences maintaining that positive incentives are more effective than negative incentives. It is certainly possible that efforts to build capacity and to grant transition periods occur only in the most difficult cases, so that our results are attributable to selection bias. Even so, we need to explore more systematically the sources of the conclusion that the data in the IRD offer no support for expectations derived from the responsiveness perspective on compliance. One of the reasons may well be that these two variables cover only part of this theoretical perspective. Our results therefore should not lead us to a general rejection of this perspective.

## 3.4 Conclusion—Compliance Strategies

We cannot explain compliance with international environmental rules and regulations purely in terms of the depth or shallowness of treaty obligations. Experts clearly acknowledge the causal impacts of institutional features; variations in the character of institutions clearly affect compliance rates. Over time, environmental regimes can become both more demanding and more effective in eliciting compliance. On the other hand, efforts to promote compliance through negotiations and various forms of capacity building are not sufficient to induce compliance with international rules and regulations. Most successful regimes rely on compliance mechanisms involving horizontal sanctioning and institutionalized verification procedures. More specifically, evidence drawn from the IRD indicates clearly that international environmental regimes that establish mechanisms of horizontal sanctioning and strong verification procedures do better than average with regard to compliance.

The variables emphasized by the theoretical perspective labeled *incentives* play an important role in the generation of compliance at the international level. Taken together, our findings suggest that neither the shallowness argument of Downs, Rocke, and Barsoom (1996) nor the management approach of Chayes and Chayes (1995) can explain patterns of compliance with international environmental regimes. In our view, a composite perspective that integrates "incentives," "institutional design," "the rule of law," and "the power of legitimacy" is needed.

Although rational institutionalism's emphasis on incentives is justified, *legalization* can add stability to compliance rates. Legalization as a concept encompasses juridification and internalization. Of the two, however, evidence from the IRD indicates that juridification is more important with regard to international environmental regimes. As the *legitimacy* perspective suggests, drawing transnational NGOs into the decision-making process helps as well. When the addresses and (possibly) those affected by rules and regulations are involved in the rule making, compliance with international environmental regimes improves.

A striking conclusion of this analysis is that *responsive* mechanisms for eliciting compliance are much less important than mechanisms associated with the other theoretical perspectives, at least with regard to international environmental regimes. The data included in the IRD do not confirm expectations about the role of capacity building. The IRD provides only indirect evidence that responsiveness plays a role in responding to specific violations, and the results indicate only moderate success.

The role that legitimacy plays in increasing compliance with international rules and regulations is especially interesting. A closer examination of this perspective suggests that legitimacy is more than an incremental determinant of compliance. A related study has shown that the influence of legitimacy is especially important when other compliance mechanisms are also strong and effective (Zürn and Neyer 2005). Combined with mechanisms based on rational institutionalism, juridification often works well and explains high rates of compliance with specific rules and regulations. The combination of these mechanisms may run into trouble, however, when an issue reaches a broader public *agenda* and when different national public discourses are both *fragmented* in the sense that they do not relate to each other and *polarized* in the sense that they lead to different outcomes. The BSE crises in the European Union and the conflicts about growth hormones in the WTO are prominent examples (Neyer 2004). Despite the occurrence of good overall compliance rates, these regulations experienced compliance crises. National politicians who were willing in principle to comply with the rules came under intense pressure from domestic interests backed by more or less homogeneous public opinion.

As long as structurally similar issues do not become a focus of public debate (as in the case of subsidy controls—Wolf 2005), by contrast, compliance is not a problem. At worst, some initial cases of noncompliance may occur. But even well-designed mechanisms for attaining compliance with intergovernmental agreements can engender compliance crises when there is no supportive public discourse among those affected by the rules. In general, the links between *legitimacy* and compliance become most important when heavily legalized rules and regulations come under stress.

The same study also concludes that the formulation of capacity-sensitive rules is relatively unimportant as a source of compliance. But this study does find evidence that the second aspect of the management approach (i.e., the degree to which the application of rules features flexibility) does make a difference. An important reason for the emergence of compliance problems is a perception on the part of subjects that they are not given sufficient opportunities to inject their concerns into the formulation of rules in such a way as to clarify their changing needs and wants. This finding—coupled with the evidence presented in this chapter that compliance improves over time—points to the operation of a form of responsiveness. This is not responsiveness treated as a matter of dealing with specific breaches in a flexible manner. Rather, it is responsiveness in the form of an ongoing discourse about compliance and the development of mechanisms to improve compliance. When regimes allow for easy adjustments to accommodate new developments, compliance improves.

Adding up all these findings yields the following conclusion. Adequate and even impressive rates of compliance with international environmental rules occur when appropriate incentive mechanisms are coupled with juridification, participation on the part of transnational NGOs in the rule-making process, and a responsive approach to the development of compliance mechanisms over time. In short, the formula for achieving compliance is: horizontal enforcement + strong verification mechanisms + juridification + NGO participation + institutional development = high rates of compliance.

# 4

# Decision Rules, Compliance Mechanisms, and the Effectiveness of Regimes

## 4.1 Introduction

Among the concerns most often articulated by those who are skeptical about the capacity of regimes to meet the demand for governance in international society, two stand out as particularly worthy of consideration: (1) limitations arising from the pressure to arrive at collective decisions by consensus, and (2) constraints attributable to the lack of capacity to enforce collective decisions once made and accepted in principle. These concerns raise fundamental questions about the effectiveness of international regimes. But should they lead us to conclude that regimes are largely epiphenomena incapable of operating effectively as sources of governance under any but the most limited of circumstances (Strange 1983; Mearsheimer 1994–1995)? In this chapter, we seek to answer this question, first by describing current practice regarding these matters and then by examining theoretical arguments pertaining to decision rules and compliance mechanisms and by making use of the International Regimes Database (IRD) to explore the empirical evidence pertaining to our theoretical expectations.

We proceed as follows. The first substantive section of the chapter focuses on the links between decision rules and regime effectiveness. Individual subsections describe current practice, explore theoretical expectations about these links, and compare these expectations to empirical evidence drawn from the IRD. The next substantive section presents a similar analysis regarding compliance mechanisms, with individual subsections again characterizing current practice, presenting theoretical arguments, and evaluating empirical evidence contained in the IRD. We have deliberately restricted the analysis in the empirical subsections to a small number of relatively simple or straightforward queries. Even so, the patterns that emerge regarding both decision rules and compliance mechanisms—at least when it comes to environmental arrangements—turn out to be considerably more complex than one might anticipate from a consideration of the theoretical arguments alone. The concluding section, then, returns to the problems identified in the opening paragraph of the chapter and includes a brief examination of the implications of our analysis of decision rules and compliance mechanisms for the (re)design of regimes dealing with environmental concerns. We cannot offer any simple prescriptions or surefire recipes for success that are applicable to all situations. But we can provide a number of insights that should be of interest to those responsible for the creation of new regimes or the reconstruction of existing institutional arrangements.

## 4.2 Decision Rules and Regime Effectiveness

What rules do international regimes employ to arrive at collective choices regarding matters that fall within their jurisdiction? Are there clearcut links between the nature of the decision rules embedded in regimes and their effectiveness, treated both as the attainment of goals and as progress toward solving the problems leading to their creation? We start with a brief account of current practices relating to these matters, using arrangements dealing with large-scale environmental matters as a source of illustrations. An analysis of the implications of these practices for effectiveness in both theoretical and empirical terms follows in the succeeding subsections.

### 4.2.1 Decision Rules—Current Practice

Decision rules stipulate conditions that must be met in order to arrive at valid collective decisions or social choices relating to issues falling within the competence of specific institutional arrangements. A simple majority rule, for example, allows a legitimate or authoritative decision to be reached with the consent of 50 percent plus one of those eligible to participate in the process. A unanimity rule, by contrast, makes explicit agreement on the part of all those eligible to participate a requirement for arriving at collective choices that are legitimate or authoritative. Phrased in this way, the concept of decision rules seems relatively simple and straightforward. Yet a brief discussion will suffice to demonstrate that this appearance of simplicity is deceptive. As the examples we introduce in the following paragraphs make clear, the complications associated with the concept of decision rules are just as prominent in connection with international regimes as they are in connection with legislative practices operating at the domestic level.[1]

Some variations in the criteria spelling out what it takes to pass a measure are relatively familiar. Thus, majorities may be simple in the sense that they must include 50 percent plus one of the participants or qualified in the sense that they require the consent of some larger proportion of the participants. Two-thirds or three-quarters of the participants are common criteria, but anything short of unanimity can be treated as a form of qualified majority. Similarly, decision rules calling for consensus are to be differentiated from those explicitly requiring unanimity. Whereas unanimity means that all participants must accept a measure explicitly, consensus simply means that no participant feels so strongly about a measure that it is prepared to go on record publicly and formally as opposing its adoption.

Analysts often assume that consensus or even unanimity is the standard or "normal" decision rule used in international governance systems (Underdal 2002), and there are in fact regimes whose constitutive documents specify such a rule. The Antarctic Treaty of 1959, for example, states that measures dealing with "matters of common interest pertaining to Antarctica" (Article IX.1) "shall become effective when approved by all the Contracting Parties whose representatives were entitled to participate in the meetings held to consider these measures" (Article IX.4). For its part, the 1979 Geneva Convention on transboundary air pollution in Europe specifies that amendments to the convention "shall be adopted by consensus of the representatives of the Contracting Parties" (Article 12.3). But these requirements are not common to all international environmental regimes. Under the 1946 convention on the regulation of whaling, for instance, decisions of the International Whaling Commission "shall be taken by a simple majority of those members voting except that a three-fourths majority of those members voting shall be required" for actions involving amendments to what is known as the Schedule (Article 3.2). Amendments to Appendices 1 and 2 of the 1973 Washington Convention on trade in endangered species of fauna and flora "shall be adopted by a two-thirds majority of Parties present and voting" (Article 15.1.b). The provisions of the 1985 Vienna Convention for the protection of the ozone layer call on the parties to reach consensus on the terms of proposed amendments. But if efforts to reach consensus fail in a particular case, the parties can adopt the amendment as a last resort "by a three-fourths majority of the Parties present and voting at the meeting" (Article 9.3).

These cases suggest as well a number of related features of decision rules that are worthy of specific attention in this discussion. To begin with, there is an important distinction between rules requiring the assent of some proportion of a regime's total membership or only of those parties present and voting. Experience with the International Whaling Commission makes it clear that situations can and do arise in which a significant number of a regime's members fail to participate, even in important decisions. Next, the criteria specified in decision rules often differ as a function of the character of the issue under consideration. The 1987 Montreal Protocol to the Vienna Convention on ozone depletion, for instance, specifies that "rules of procedure" are to be adopted by consensus (Article 11.3.a), though the regime does not require consensus in making substantive decisions on issues arising under the terms of these rules (Articles 2.9 and 2.10). And many regimes require a larger majority—or even consensus—when it comes to issues deemed particularly important than they do with respect to issues of lesser importance. One common procedure, in this regard, is to require larger majorities for actions that would change the character of the regime than for activities taking place within the existing rules of the game.

In addition, there are decision rules that establish systems of weighted voting or spell out requirements regarding the identity of some of the parties that must concur for a collective choice to be deemed legitimate or authoritative. Although weighted voting is not common in mainstream environmental regimes, it is a prominent feature of other arrangements, such as the International Monetary Fund and the World Bank, whose actions can have far-reaching environmental impacts. A striking example of provisions relating to the identity of the parties, on the other hand, appears in the 1990 London Amendments to the Montreal Protocol, which allow for decisions of the Montreal Protocol Multilateral Fund to be taken by a two-thirds majority vote of parties present and voting, but only when that majority includes a simple majority both of the developed countries and the developing countries present and voting (Article 10.9).

The identity and composition of the decision-making body may also make a difference when it comes to the operation of decision rules. In the case of Antarctica, for example, the combination of the distinction between Consultative Parties and others and the practice of making decisions about the operation of the regime at Antarctic Treaty Consultative Meetings (ATCMs) has meant that the requirement for unanimity has applied in practice only to the Consultative Parties. If it had entered into force, by contrast, the 1988 Antarctic minerals convention would have established an overall Commission authorized to "take decisions on matters of substance by a three-quarters majority of the members present and voting" (Article 22.1) as well as a set of Regulatory Committees—one for each area opened for exploration—authorized to take decisions by "a two-thirds majority of the members present and voting" (Article 32.1).[2] The fact that this convention failed to garner the support needed to enter into force has had the effect of maintaining the relative simplicity of the Antarctic Treaty System's decision rules. In the case of ozone depletion, there is a Conference of the Parties (COP) of the Vienna Convention and a Meeting of the Parties (MOP) of the Montreal Protocol, which meet concomitantly but which have memberships that are not identical and are subject to the specific provisions of their respective constitutive agreements. What is more, the Executive Committee of the Montreal Protocol Multilateral Fund established under Article X of the London Amendments is quite distinct from the COP and the MOP and operates under a decision rule that differs from that of its parent bodies.

A distinctive feature of the decision rules associated with international regimes concerns the status of measures once the Consultative Meeting, Conference of the Parties, Commission, or other responsible body has approved them. Whereas ratification by individual parties is required for some measures to take effect, no such requirement is imposed in other cases. In the case of ozone depletion, for example, the addition of new chemicals or families of chemicals—like carbon tetrachloride and methyl chloroform, referred to in Article II of the London Amendments—to the list of controlled substances requires ratification on the part of individual members to become effective. But this is not the case with respect to changes in the phaseout schedules for chemicals—like many chlorofluorocarbons and halons—already included in the list of controlled substances. Equally important is the opportunity provided in some environmental regimes for individual members to file objections or reservations, which effectively make the regime nonbinding as far as they are concerned with regard to the particular issue(s) at stake. Amendments to the Schedule of the whaling convention, for instance, "shall not become effective with respect to any Government which has... objected until such date as the objection is withdrawn" (Article V.3). Similarly, CITES allows individual parties to enter reservations with regard to amendments to Appendices I and II and specifies that "until such reservation is withdrawn, the Party shall be treated as a State not a party to the present Convention with respect to trade in the species concerned" (Article XV.3). Interestingly, the 1985 Vienna Convention on ozone, which is widely regarded as a weak agreement, states simply that "no reservations may be made to this Convention" (Article 18). Under the circumstances, the issue of whether to allow individual parties to opt in or out with regard to specific decisions arrived at under the terms of a regime becomes a matter of concern at the international level that has no direct analog at the domestic level.

This discussion does not begin to exhaust the set of distinctions regarding decision rules that are worthy of consideration in thinking about the performance of international environmental regimes. But it is surely sufficient to demonstrate the validity of the point made at the beginning of this section regarding the diversity and the complexity of decision rules as they evolve under real-world conditions. Obviously, this complexity must be taken into account in any analysis of the links between decision rules and regime effectiveness. Yet the existence of the range of decision rules described in the preceding paragraphs makes it clear that there is substantial variation among regimes in this realm. It should therefore be possible to initiate both analytic and empirical assessments of the links between decision rules and effectiveness in this domain.

### 4.2.2 Decision Rules—Theoretical Expectations

What can theory tell us about the links between decision rules on the one hand and the effectiveness of environmental regimes on the other? Here, we examine the principal arguments in the theoretical literature that are pertinent to this question. As it turns out, there are theoretical reasons to support a number of different expectations regarding the nature of these links. We single out and comment on the most important of these reasons, noting that they are associated in many cases with the larger debate about the relative merits of collective-action models and social-practice models of regimes and their consequences (Young 2002c). This will set the stage for the examination in the following subsection of empirical evidence drawn from the International Regimes Database that bears on these matters.

The most prominent debate in the literature on the links between decision rules and regime effectiveness centers on the consequences of relying on consensus or even unanimity rules in contrast to majority rules of one sort or another. Two distinct streams of thought regarding these links are particularly worthy of consideration. One focuses specifically on incentives and contrasts the rising transaction costs associated with movement in the direction of unanimity with growing losses of individual welfare arising from the use of more hierarchical decision rules. The other, rooted in the social-practice perspective on institutions, directs attention to issues of legitimacy and takes the view that regime members who acknowledge the legitimacy of a regime will not be concerned about the exact nature of the decision rule it employs. Our discussion of these theoretical matters culminates in an explicit formulation of the expectations they generate regarding what we are likely to discover in our examination of the empirical evidence in the next subsection.

Turn first to the utilitarian issues involving transaction costs and losses of welfare. Noting that the "decision rule is an important determinant of the capacity of an institution to aggregate diverging preferences," Underdal (2002, 24) observes that "aggregation capacity reaches its maximum in strictly hierarchical structures, and is at its lowest in systems requiring unanimity." A unanimity rule grants each regime member veto power over collective choices and leads "to the 'law of the least ambitious program,' meaning that collective action will be limited to those measures that are acceptable to the least enthusiastic party" (Underdal 2002, 25). Under the circumstances a shift away from unanimity should increase institutional capacity and, in the process, enhance the effectiveness of a regime in solving the problem(s) that led to its creation.

How far does this proposition hold? Is there a point at which movement away from the pole of unanimity and toward the pole of strict hierarchy will no longer yield increases in the effectiveness of regimes? Buchanan and Tullock address this issue in generic terms, comparing increases in transaction costs arising from movement toward the pole of unanimity with welfare losses to those whose preferences are ignored as a result of movement toward strict hierarchy (Buchanan and Tullock 1962). Considering the marginal costs associated with such changes, they argue that there should be an optimal point at which marginal reductions in transaction costs just equal marginal increases in the loss of individual welfare.[3] The idea that this point constitutes the locus of the optimal decision rule is a normative prescription rather than an empirical prediction. Nonetheless, this line of thinking does suggest that actors creating international regimes will typically select some sort of majority rule—simple or qualified as the case may be.

Is this generic argument compelling at the international level? International society differs from other social systems due to the decentralized character of its political system and its underdeveloped capacity to enforce collective decisions accepted in principle. Members of international regimes are unusually zealous in guarding their sovereign right to refuse to be bound by choices they have not accepted explicitly, and they exploit various devices (e.g., filing reservations or objections) that allow them to opt out of decisions they do not like. What this implies is that many regime members would rather accept high transaction costs and even the prospect of paralysis than run the risk of welfare losses arising from the use of some sort of majority rule. This suggests, at a minimum, that the effectiveness of collective choices at the international level is apt to depend heavily on the cultivation of voluntary acceptance or, in other words, some form of consent on the part of individual regime members.

Nor is it obvious that reliance on a de facto consensus rule—not to be confused with unanimity—will produce results that are less effective than outcomes arising from rules calling for some sort of qualified majority.

This is especially true when a consensus rule is coupled with other arrangements, such as procedures that ignore the views of those choosing to abstain and that provide opportunities for those with serious concerns to protect their interests even while going along with the consensus (e.g., exploiting loopholes like the provision allowing "scientific whaling" in the regime for whales and whaling). Unlike unanimity or some specified majority, consensus turns out to be a rather elastic decision rule. It can range all the way from situations featuring universal enthusiasm to situations in which some parties accept collective choices grudgingly and only because they prefer to avoid the political fallout likely to result from open opposition. What is more, a decision adopted by consensus, which does not trigger the filing of formal objections, may emerge later on as an instrument that can be used to bring pressure to bear on reluctant parties to adjust their behavior to conform to the terms of the consensus.

Contrast this line of analysis with expectations about decision rules arising from a social-practice perspective on the consequences of such rules. This perspective has several elements that have significant implications for a discussion of the links between decision rules and the effectiveness of regimes. First, and perhaps foremost, is the idea rooted in the agent-structure debate that institutions play important roles in shaping the identity of their members and, in the process, determining the content of their preference systems. Rather than treating regime members as actors with preexisting and firmly established preference systems who come together to negotiate a social contract, this way of thinking suggests that membership in various institutional arrangements is a significant force in shaping the preferences of regime members (Wendt 1999). The implication of this view is that while individual members will not only accept the provisions of regimes without engaging in utilitarian calculations, they will also exhibit a pronounced propensity to regard collective choices arising from the operation of a regime as legitimate without regard to the formal character of the decision rule embedded in the provisions of the regime.

This is not equivalent to the argument spelled out above regarding the role of consensus. In contrast to the logic underlying the idea of minimum winning coalitions, the pursuit of consensus combines an effort to maximize the number of parties willing to accept a particular decision with an acknowledgment of the importance of escape routes that allow dissident members to protect their interests without blocking the will of the majority. The social-practice perspective, by contrast, suggests that the exact nature of the decision rule in use under the terms of any given regime may not make much difference with regard to the effectiveness of the regime. So long as individual members regard the provisions of a regime as legitimate and accept the results it generates as authoritative, effectiveness will not vary as a function of the precise nature of the decision rule used to arrive at collective choices. Even arrangements that have no explicit decision rule may prove effective on this account. Within limits, then, this perspective suggests that we should expect to find no clearcut relationship between decision rules and regime effectiveness.

Before leaving this theoretical account, let us recap the main points of the discussion and crystallize the resultant expectations about what we should expect to find when we turn to data drawn from the IRD. Those whose thinking is rooted in collective-action models argue that unanimity rules are apt to be avoided because they drive up transaction costs and, in the process, activate the law of the least ambitious program. At the same time, their assumptions about factors that limit the prospects for enforcing compliance produce the conclusion that the use of consensus rules made flexible through the inclusion of suitable escape clauses will maximize the effectiveness of international environmental regimes. Those who think in social-practice terms, on the other hand, expect to find no discernible relationship between the character of a regime's decision rules and the effectiveness of the regime in solving or managing environmental problems. They assume, in essence, that members who acknowledge the legitimacy of the regime itself will accept the choices arising from the operation of its decision-making procedures, regardless of the character of the decision rule employed.

### 4.2.3 Decision Rules—Empirical Evidence

The ideas reviewed in the preceding subsection are analytic in character. They are rooted in theories or models that seek to reveal the ways in which alternative decision rules will influence the behavior of various actors and ultimately contribute to solving the problems that gave rise to regime formation. This is a valuable contribution. But it stops short of comparing these analytic expectations with empirical evidence regarding the operation of environmental regimes in a variety of real-world settings. To fill this gap, we turn now to an examination of data included in the International Regimes Database that can illuminate the links between decision rules and the effectiveness of international environmental regimes. The fact that we have identified competing hypotheses or expectations about these links arising from collective-action models and social-practice models makes the shift to empirical analysis particularly important. Although we do not expect this exercise to yield definitive results regarding the relative merits of the two families of models, we do anticipate that drawing on the IRD's data will help to sharpen and deepen our understanding of the behavioral mechanisms through which regimes guide the course of interactive behavior at the international level.

Part II of the IRD Data Protocol, which focuses on regime attributes, contains a number of variables relating to decision making. In the analysis to follow, we have chosen to work with two of these variables called "decision rules provided" and "decision rules in practice." The first of these variables gives rise to the question: "What decision rules does the regime provide for in arriving at decisions?" The second revises this question to ask: "What decision rules does the regime use in practice in arriving at decisions?" In both cases, the menu of answers includes: no decision rules, unanimity, consensus, weighted or unweighted voting, qualified majority, simple majority, and right to opt out or file objections. This allows the IRD to provide data that can lead to insights regarding the familiar conceptual distinction between rules on paper and rules in practice.

If we approach these considerations regarding decision rules as independent variables, it makes sense to look for measures of regime effectiveness in the IRD to be paired with them as dependent variables. Part III of the IRD, which deals with regime consequences, contains a number of questions pertaining to effectiveness, many of which are of interest from the perspective of probing the theoretical concerns outlined in the preceding subsection. To make the analysis manageable, however, we have chosen to focus on matters of goal attainment and problem change as dependent variables.

In the case of goal attainment, there is an important distinction between goal attainment as such and the causal roles that regimes play in bringing about goal attainment. Thus, the IRD starts by asking coders the following question: "If the regime has stated goals, indicate whether the behavioral changes led to the fulfillment of the stated goals of the regime." The next query then asks: "For each stated goal coded under GOALS_FULFILL, indicate what causal impact the regime had on this development." As a result, we seek not only to move beyond regime outputs and outcomes to examine impacts but also to draw a distinction between the causal role of the regime itself and other driving forces relevant to goal attainment. Similar observations are in order regarding the issue of problem solving. Thus, the IRD starts with the following request: "For each problem listed under PROBLEM, indicate whether and how the state of the world changed during this period with respect to the problem." A follow-up query then asks: "For each problem coded under PROBLEM_CHANGE indicate whether the regime exerted a causal influence on the change of the world with regard to the problem."

We can proceed now to construct a set of select queries in MS Access that makes it possible to interrogate the IRD in order to shed light on our theoretical expectations regarding the links between decision rules and effectiveness. To be specific, we have constructed four queries dealing with these links:

*Query 1* Are there clearcut links between the decision rules as provided in constitutive documents and goal attainment?

*Query 2* Are there clearcut links between decision rules in practice and goal attainment?

*Query 3* Are there identifiable connections between decision rules as provided in constitutive documents and problem change?

*Query 4* Are there identifiable connections between decision rules in practice and problem change?

What can we learn from the results of these queries, and what light do they shed on the theoretical expectations articulated in the previous subsection? Turn first to tables 4.1 and 4.2, dealing with the relationship between decision rules and goal attainment.

A finding that jumps out immediately is that the decision rules in use among environmental regimes are far more likely to feature consensual procedures than an examination of decision rules on paper would lead one to expect. While the constitutive documents call for consensus in 20.2 percent of the records, procedures featuring consensus emerge as the dominant method of reaching decisions in 58.4 percent of the records. Whatever the nature of the decision rules agreed to during the stage of institutional bargaining leading to regime formation, in other words, those charged with administering or operating regimes resort in practice to the use of consensual procedures more than half the time.

The data dealing with rules on paper show that constitutive documents call for decision making via unanimity 16 percent of the time and for the use of rules featuring some form of qualified majority 22.1 percent of the time. There is no explicit decision rule built into the regimes included in the database in 14.1 percent of the records. But the data on decision rules in practice tell quite a different story. Beyond the preponderance of cases using consensual procedures, unanimity rules are employed in practice in 21.9 percent of the records. All the other decision rules are used in less than 10 percent of the cases. What can we infer from these data? They are compatible—as those who think in collective-action terms would expect—with a strong tendency to rely on consensual procedures in efforts to ensure that regimes matter or, in other words, that subjects will comply with the decisions made under the terms of these arrangements. Nonetheless, the fact that regimes in practice still make use of unanimity rules in 21.9 percent of the cases makes it clear that sovereignty sensitivity with its emphasis on the importance of explicit agreement on the part of individual member states is still very much alive.

A number of decision rules are associated with institutional arrangements that perform well in terms of goal attainment. Examining the data on decision rules in practice included in table 4.2, we can see that regimes relying on unanimity rules fulfill their goals in 87.3 percent of the cases in the database. Other rules are associated with arrangements that do nearly as well in these terms. Qualified majorities are used in arrangements in which goals are fulfilled 78.9 percent of the time. The

**Table 4.1**
Decision rules provided and goal attainment

| Attributes | Count | | Goals fulfilled | | Degree of Causality_Goals fulfilled | | | | | | Degree of Causality_Goals not fulfilled | | | | | |
|---|---|---|---|---|---|---|---|---|---|---|---|---|---|---|---|---|
| | | | | | Little or no causal impact | | Modest causal influence | | Large causal influence | | Little or no causal impact | | Modest causal influence | | Large causal influence | |
| No decision rules | 233 | 14.26% | Yes | | **4** | 2.34% | **11** | 6.43% | **156** | 91.23% | **16** | 25.81% | **46** | 74.19% | **0** | 0% |
| | | | 171 | 73.39% | | | | | | | | | | | | |
| | | | No | | | | | | | | | | | | | |
| | | | 62 | 26.61% | | | | | | | | | | | | |
| Unanimity | 443 | 27.11% | Yes | | **16** | 4.65% | **155** | 45.06% | **173** | 50.29% | **12** | 12.12% | **26** | 26.26% | **61** | 61.62% |
| | | | 344 | 77.65% | | | | | | | | | | | | |
| | | | No | | | | | | | | | | | | | |
| | | | 99 | 22.35% | | | | | | | | | | | | |
| Consensus | 330 | 20.19% | Yes | | **3** | 1.34% | **63** | 28.13% | **158** | 70.54% | **53** | 50.00% | **45** | 42.45% | **8** | 7.55% |
| | | | 224 | 67.88% | | | | | | | | | | | | |
| | | | No | | | | | | | | | | | | | |
| | | | 106 | 32.12% | | | | | | | | | | | | |

| | | | Yes | | No | | | | | | | | | | | | | |
|---|---|---|---|---|---|---|---|---|---|---|---|---|---|---|---|---|---|---|
| Weighted/ unweighted voting | 47 | 2.88% | 25 | 53.19% | 22 | 46.81% | 0 | 0% | 25 | 100% | 0 | 0% | 10 | 45.45% | 4 | 18.18% | 8 | 36.36% |
| Qualified majority | 360 | 22.03% | 265 | 73.61% | 95 | 26.39% | 30 | 11.32% | 29 | 10.94% | 206 | 77.74% | 50 | 52.63% | 38 | 40.00% | 7 | 7.37% |
| Simple majority | 55 | 3.37% | 42 | 76.36% | 13 | 23.64% | 5 | 11.90% | 15 | 35.71% | 22 | 52.38% | 6 | 46.15% | 7 | 53.85% | 0 | 0% |
| Right to opt-out, file objection | 166 | 10.16% | 123 | 75.46% | 43 | 24.54% | 23 | 18.70% | 21 | 17.07% | 79 | 64.23% | 18 | 41.86% | 13 | 30.23% | 12 | 27.91% |
| Totals | 1634 | 100.00% | 1194 | 73.07% | 440 | 26.93% | 81 | | 319 | | 794 | | 165 | | 179 | | 96 | |

**Table 4.2**
Decision rules in practice and goal attainment

| Attributes | Count | % | Goals fulfilled | | | Degree of Causality_Goals fulfilled | | | | | | Degree of Causality_Goals not fulfilled | | | | | |
|---|---|---|---|---|---|---|---|---|---|---|---|---|---|---|---|---|---|
| | | | | | | Little or no causal impact | | Modest causal influence | | Large causal influence | | Little or no causal impact | | Modest causal influence | | Large causal influence | |
| No decision rules | 43 | 3.33% | **Yes** | 23 | 53.49% | **0** | 0.00% | **6** | 26.09% | **17** | 73.91% | **3** | 15.00% | **17** | 85.00% | **0** | 0% |
| | | | **No** | 20 | 46.51% | | | | | | | | | | | | |
| Unanimity | 275 | 21.30% | **Yes** | 240 | 87.27% | **14** | 5.83% | **144** | 60.00% | **82** | 34.17% | **6** | 17.14% | **25** | 71.43% | **4** | 11.43% |
| | | | **No** | 35 | 12.73% | | | | | | | | | | | | |
| Consensus | 754 | 58.40% | **Yes** | 587 | 77.85% | **34** | 5.79% | **119** | 20.27% | **434** | 73.94% | **93** | 55.69% | **62** | 37.13% | **12** | 7.19% |
| | | | **No** | 167 | 22.15% | | | | | | | | | | | | |

| | | | | | | | | | | | | | | | | | | |
|---|---|---|---|---|---|---|---|---|---|---|---|---|---|---|---|---|---|---|
| Weighted/ unweighted voting | 5 | 0.39% | Yes | | | 0 | 0% | 1 | 100% | 0 | 0% | 0 | 0.00% | 4 | 100.00% | 0 | 0.00% |
| | | | 1 | 20.00% | | | | | | | | | | | | | |
| | | | No | | | | | | | | | | | | | | |
| | | | 4 | 80.00% | | | | | | | | | | | | | |
| Qualified majority | 104 | 8.06% | Yes | | | 7 | 8.54% | 18 | 21.95% | 57 | 69.51% | 6 | 27.27% | 10 | 45.45% | 6 | 27.27% |
| | | | 82 | 78.85% | | | | | | | | | | | | | |
| | | | No | | | | | | | | | | | | | | |
| | | | 22 | 21.15% | | | | | | | | | | | | | |
| Simple majority | 15 | 1.16% | Yes | | | 0 | 0.00% | 7 | 46.67% | 8 | 53.33% | 0 | 0.00% | 0 | 0.00% | 0 | 0.00% |
| | | | 15 | 100.00% | | | | | | | | | | | | | |
| | | | No | | | | | | | | | | | | | | |
| | | | 0 | 0.00% | | | | | | | | | | | | | |
| Right to opt-out, file objection | 95 | 7.36% | Yes | | | 13 | 19.40% | 21 | 31.34% | 33 | 49.25% | 12 | 42.86% | 10 | 35.71% | 6 | 21.43% |
| | | | 67 | 70.53% | | | | | | | | | | | | | |
| | | | No | | | | | | | | | | | | | | |
| | | | 28 | 29.47% | | | | | | | | | | | | | |
| **Totals** | **1291** | **100%** | | | | 68 | | 316 | | 631 | | 120 | | 168 | | 28 | |

comparable figures for systems using consensual procedures and for systems allowing for opt-out procedures are 77.9 and 70.5 percent respectively. Even systems that have no explicit decision rules achieve goal attainment in 53.5 percent of the records. Thus, regimes fulfill their goals in a large proportion of the cases, and they do so using a variety of decision rules. This may be interpreted as evidence in favor of the social-practice perspective, leading us to expect no clearcut relationship between the type of decision rules employed and the prospects for achieving success in efforts to address environmental problems.

Note, however that the data reported in the previous paragraph do not tell us anything about the extent to which a regime plays a causal role in pursuing the objective of goal attainment. It is perfectly possible, in most cases, that goal attainment is attributable to factors other than the existence and operation of the regimes themselves. What do the data have to say about this prospect? In general, the regimes included in the database are assessed as having a large causal influence in a sizable proportion of the cases. Thus, regimes that have no explicit decision rules still have a large causal influence in the pursuit of goal attainment in 73.9 percent of the records. Comparable figures for consensus and qualified majority voting are 73.8 and 69.5 percent respectively. One interesting finding in this context is that regimes relying on unanimity as a decision rule produce a large causal influence in the pursuit of goal attainment only 34.2 percent of the time. It appears, then, that the use of consensual procedures or even the absence of an explicit decision rule produces the best results in terms of causal impact. It is worth emphasizing here that regimes leading to goal attainment are rated as producing either a large causal influence or a modest causal influence in most cases. In cases where regimes fail to produce goal attainment, on the other hand, they typically rely on decision rules featuring qualified majorities, opt-out procedures, or the requirement of unanimity. None of this yields simple generalizations regarding the relationship between decision rules and success construed as a matter of goal attainment. Still, it seems undeniable that regimes matter when it comes to the pursuit of goal attainment. And there is much to be said for the arguments that it makes sense to use consensual procedures when explicit decision rules are needed and that there is a class of situations in which it does no harm to refrain from spelling out any decision rule at all in explicit terms. These observations are subject to plausible interpretations that are compatible with either a collective-choice perspective or a social-practice perspective.

It is obvious, of course, that goal attainment is by no means the only perspective on regime effectiveness or the success of these institutional arrangements. In fact, there are cases in which goals—or at least those goals stated publicly—are quite modest, so that they can be fulfilled without making much of a dent in the underlying problem. For those focused on problem solving, the real issue is whether the problem got better or worse following the creation of a regime rather than the extent to which a regime plays a demonstrable role in fulfilling stated goals. In the nature of things, identifying the causal pathways leading toward or away from problem solving is a difficult task. But the IRD does contain data that allow us to ask and answer questions regarding the relationship between decision rules and progress toward solving the problems that lead decision makers to create them in the first place. These data are reported in tables 4.3 and 4.4.

Because the queries giving rise to these data differ from the queries producing data on goal attainment, we start again with some descriptive observations about the incidence of distinct decision rules. Broadly speaking, the results are compatible with those reported in the tables pertaining to goal attainment. On paper, 25.6 percent of the cases call for the use of a unanimity rule, 23 percent call for consensual procedures, and 21 percent rely on some form of qualified majority. But once again, the picture is quite different when we turn to decision rules in use in contrast to rules on paper. Some 53.8 percent of the cases rely in practice on consensual procedures, whereas 20.5 percent still rely on unanimity. The next most common decision rule, opt-out procedures, appears in only 11.9 percent of the records. As in the case of goal attainment, we can infer that there are strong incentives among those responsible for administering or operating regimes to turn to consensual procedures, regardless of what the formal provisions of constitutive documents may say.

What can we say about the relationship between the decision rules employed in practice and the extent to which the relevant problems improve or worsen? In this case, we have a five-point ordinal scale ranging from improved considerably through improved slightly, stayed the same,

**Table 4.3**
Decision rules provided and problem change

| Attributes | Count | % | Problem change | | Problem change causal | | | | | | | | | |
|---|---|---|---|---|---|---|---|---|---|---|---|---|---|---|
| | | | | | Little or no causal impact | | Modest causal impact | | Balanced causal influence | | Significant causal influence | | Very strong causal influence | |
| No decision rules | 62 | 9.44% | worsened considerably | | 0 | 0% | 4 | 100% | 0 | 0% | 0 | 0% | 0 | 0% |
| | | | 4 | 6.45% | | | | | | | | | | |
| | | | worsened slightly | | 0 | 0% | 7 | 77.78% | 0 | 0% | 2 | 22.22% | 0 | 0% |
| | | | 9 | 14.52% | | | | | | | | | | |
| | | | stayed the same | | 2 | 11.76% | 5 | 29.41% | 0 | 0% | 0 | 0% | 10 | 58.82% |
| | | | 17 | 27.42% | | | | | | | | | | |
| | | | improved slightly | | 7 | 33.33% | 0 | 0% | 2 | 9.52% | 12 | 57.14% | 0 | 0% |
| | | | 21 | 33.87% | | | | | | | | | | |
| | | | improved considerably | | 0 | 0% | 0 | 0% | 0 | 0% | 9 | 81.82% | 2 | 18.18% |
| | | | 11 | 17.74% | | | | | | | | | | |

| | | | | | | | | | | | | | | |
|---|---|---|---|---|---|---|---|---|---|---|---|---|---|---|
| Unanimity | 168 | 25.57% | worsened considerably | | 2 | 40% | 2 | 40% | 0 | 0% | 1 | 20% | 0 | 0% |
| | | | 5 | 2.98% | | | | | | | | | | |
| | | | worsened slightly | | 28 | 60.87% | 12 | 26.09% | 4 | 8.70% | 2 | 4.35% | 0 | 0% |
| | | | 46 | 27.38% | | | | | | | | | | |
| | | | stayed the same | | 0 | 0% | 12 | 37.50% | 2 | 6.25% | 18 | 56.25% | 0 | 0% |
| | | | 32 | 19.05% | | | | | | | | | | |
| | | | improved slightly | | 11 | 37.93% | 5 | 17.24% | 5 | 17.24% | 8 | 27.59% | 0 | 0% |
| | | | 29 | 17.26% | | | | | | | | | | |
| | | | improved considerably | | 31 | 55.36% | 2 | 3.57% | 8 | 14.29% | 9 | 16.07% | 6 | 10.71% |
| | | | 56 | 33.33% | | | | | | | | | | |

**Table 4.3**
(continued)

| Attributes | Count | % | Problem change | | Problem change causal | | | | | | | | | |
|---|---|---|---|---|---|---|---|---|---|---|---|---|---|---|
| | | | | | Little or no causal impact | | Modest causal impact | | Balanced causal influence | | Significant causal influence | | Very strong causal influence | |
| Consensus | 150 | 22.83% | **worsened considerably** | | 5 | 29.41% | 10 | 58.82% | 0 | 0% | 0 | 0% | 2 | 11.76% |
| | | | 17 | 11.33% | | | | | | | | | | |
| | | | **worsened slightly** | | 6 | 30.00% | 5 | 25.00% | 4 | 20.00% | 5 | 25.00% | 0 | 0% |
| | | | 20 | 13.33% | | | | | | | | | | |
| | | | **stayed the same** | | 15 | 50% | 4 | 13.33% | 2 | 6.67% | 3 | 10% | 6 | 20% |
| | | | 30 | 20% | | | | | | | | | | |
| | | | **improved slightly** | | 2 | 4.08% | 11 | 22.45% | 10 | 20.41% | 7 | 14.29% | 19 | 38.78% |
| | | | 49 | 32.67% | | | | | | | | | | |
| | | | **improved considerably** | | 0 | 0% | 0 | 0% | 12 | 35.29% | 12 | 35.29% | 10 | 29.41% |
| | | | 34 | 22.67% | | | | | | | | | | |

| | | | | | | | | | | | | | | |
|---|---|---|---|---|---|---|---|---|---|---|---|---|---|---|
| Weighted/ unweighted voting | 19 | 2.89% | worsened considerably | | 1 | 11.11% | 0 | 0% | 0 | 0% | 8 | 88.89% | 0 | 0% |
| | | | 9 | 47.37% | | | | | | | | | | |
| | | | worsened slightly | | 1 | 100% | 0 | 0% | 0 | 0% | 0 | 0% | 0 | 0% |
| | | | 1 | 5.26% | | | | | | | | | | |
| | | | stayed the same | | 1 | 100% | 0 | 0% | 0 | 0% | 0 | 0% | 0 | 0% |
| | | | 1 | 5.26% | | | | | | | | | | |
| | | | improved slightly | | 0 | 0% | 0 | 0% | 8 | 100% | 0 | 0% | 0 | 0% |
| | | | 8 | 42.11% | | | | | | | | | | |
| | | | improved considerably | | 0 | 0% | 0 | 0% | 0 | 0% | 0 | 0% | 0 | 0% |
| | | | 0 | 0.00% | | | | | | | | | | |

**Table 4.3**
(continued)

| Attributes | Count | % | Problem change | | | Problem change causal | | | | | | | | | |
|---|---|---|---|---|---|---|---|---|---|---|---|---|---|---|---|
| | | | | | | Little or no causal impact | | Modest causal impact | | Balanced causal influence | | Significant causal influence | | Very strong causal influence | |
| Qualified majority | 144 | 21.92% | worsened considerably | 30 | 20.83% | 9 | 30% | 19 | 63.33% | 0 | 0% | 2 | 6.67% | 0 | 0% |
| | | | worsened slightly | 32 | 22.22% | 19 | 59.38% | 13 | 40.63% | 0 | 0% | 0 | 0% | 0 | 0% |
| | | | stayed the same | 25 | 17.36% | 2 | 8% | 12 | 48% | 3 | 12% | 2 | 8% | 6 | 24% |
| | | | improved slightly | 34 | 23.61% | 7 | 21% | 5 | 15% | 10 | 29% | 12 | 35% | 0 | 0% |
| | | | improved considerably | 23 | 15.97% | 0 | 0% | 0 | 0% | 12 | 52.17% | 9 | 39.13% | 2 | 8.70% |

| | | | | | | | | | | | | | | |
|---|---|---|---|---|---|---|---|---|---|---|---|---|---|---|
| Simple majority | 21 | 3.20% | worsened considerably | | 1 | 100% | 0 | 0% | 0 | 0% | 0 | 0% | 0 | 0% |
| | | | 1 | 4.76% | | | | | | | | | | |
| | | | worsened slightly | | 1 | 50% | 1 | 50% | 0 | 0% | 0 | 0% | 0 | 0% |
| | | | 2 | 9.52% | | | | | | | | | | |
| | | | stayed the same | | 0 | 0% | 0 | 0% | 0 | 0% | 0 | 0% | 0 | 0% |
| | | | 0 | 0% | | | | | | | | | | |
| | | | improved slightly | | 1 | 8.33% | 4 | 33.33% | 5 | 41.67% | 2 | 16.67% | 0 | 0% |
| | | | 12 | 57.14% | | | | | | | | | | |
| | | | improved considerably | | 1 | 16.67% | 1 | 16.67% | 2 | 33.33% | 2 | 33.33% | 0 | 0% |
| | | | 6 | 28.57% | | | | | | | | | | |

**Table 4.3**
(continued)

| Attributes | Count | % | Problem change | | | Problem change causal | | | | | | | | | |
|---|---|---|---|---|---|---|---|---|---|---|---|---|---|---|---|
| | | | | | | Little or no causal impact | | Modest causal impact | | Balanced causal influence | | Significant causal influence | | Very strong causal influence | |
| Right to opt-out, file objection | 93 | 14.16% | **worsened considerably** | 15 | 16.13% | 5 | 33.33% | 8 | 53.33% | 0 | 0% | 2 | 13.33% | 0 | 0% |
| | | | **worsened slightly** | 34 | 36.56% | 23 | 67.65% | 5 | 14.71% | 4 | 11.76% | 2 | 5.88% | 0 | 0% |
| | | | **stayed the same** | 13 | 13.98% | 1 | 7.69% | 6 | 46.15% | 2 | 15.38% | 4 | 30.77% | 0 | 0% |
| | | | **improved slightly** | 17 | 18.28% | 3 | 17.65% | 2 | 11.76% | 4 | 23.53% | 8 | 47.06% | 0 | 0% |
| | | | **improved considerably** | 14 | 15.05% | 0 | 0% | 0 | 0% | 6 | 42.86% | 6 | 42.86% | 2 | 14.29% |
| **Totals** | 657 | 100.00% | | | | 185 | | 155 | | 105 | | 147 | | 65 | |

**Table 4.4**
Decision rules in practice and problem change

| Attributes | Count | % | Problem change | | Problem_Change_Causal | | | | | | | | | |
|---|---|---|---|---|---|---|---|---|---|---|---|---|---|---|
| | | | | | Little or no causal impact | | Modest causal impact | | Balanced causal influence | | Significant causal influence | | Very strong causal influence | |
| No decision rules | 11 | 1.98% | worsened considerably | | 0 | 0% | 0 | 0% | 0 | 0% | 0 | 0% | 0 | 0% |
| | | | 0 | 0% | | | | | | | | | | |
| | | | worsened slightly | | 0 | 0% | 0 | 0% | 0 | 0% | 2 | 100% | 0 | 0% |
| | | | 2 | 18.18% | | | | | | | | | | |
| | | | stayed the same | | 2 | 28.57% | 1 | 14.29% | 0 | 0% | 0 | 0% | 4 | 57.14% |
| | | | 7 | 63.64% | | | | | | | | | | |
| | | | improved slightly | | 1 | 50% | 0 | 0% | 1 | 50% | 0 | 0% | 0 | 0% |
| | | | 2 | 18.18% | | | | | | | | | | |
| | | | improved considerably | | 0 | 0% | 0 | 0% | 0 | 0% | 0 | 0% | 0 | 0% |
| | | | 0 | 0% | | | | | | | | | | |

**Table 4.4**
(continued)

| Attributes | Count | % | Problem change | | Problem_Change_Causal | | | | | | | | | |
|---|---|---|---|---|---|---|---|---|---|---|---|---|---|---|
| | | | | | Little or no causal impact | | Modest causal impact | | Balanced causal influence | | Significant causal influence | | Very strong causal influence | |
| Unanimity | 114 | 20.50% | **worsened considerably** | | 2 | 50% | 2 | 50% | 0 | 0% | 0 | 0% | 0 | 0% |
| | | | 4 | 3.51% | | | | | | | | | | |
| | | | **worsened slightly** | | 12 | 40.00% | 12 | 40.00% | 4 | 13.33% | 2 | 6.67% | 0 | 0% |
| | | | 30 | 26.32% | | | | | | | | | | |
| | | | **stayed the same** | | 0 | 0% | 12 | 31.58% | 2 | 5.26% | 24 | 63.16% | 0 | 0% |
| | | | 38 | 33.33% | | | | | | | | | | |
| | | | **improved slightly** | | 11 | 52.38% | 0 | 0% | 2 | 9.52% | 8 | 38.10% | 0 | 0% |
| | | | 21 | 18.42% | | | | | | | | | | |
| | | | **improved considerably** | | 0 | 0% | 3 | 14.29% | 7 | 33.33% | 5 | 23.81% | 6 | 28.57% |
| | | | 21 | 18.42% | | | | | | | | | | |

| | | | | | | | | | | | | | | | |
|---|---|---|---|---|---|---|---|---|---|---|---|---|---|---|---|
| Consensus | 299 | 53.78% | worsened considerably | 40 | 13.38% | 12 | 30.00% | 17 | 42.50% | 0 | 0% | 9 | 22.50% | 2 | 5% |
| | | | worsened slightly | 60 | 20.07% | 33 | 55.00% | 18 | 30.00% | 4 | 6.67% | 5 | 8.33% | 0 | 0% |
| | | | stayed the same | 51 | 17% | 15 | 29% | 16 | 31.37% | 3 | 5.88% | 3 | 5.88% | 14 | 27% |
| | | | improved slightly | 87 | 29.10% | 6 | 6.90% | 12 | 13.79% | 20 | 26.23% | 30 | 34.48% | 19 | 21.84% |
| | | | improved considerably | 61 | 20.40% | 2 | 3.28% | 3 | 4.92% | 13 | 21.31% | 27 | 44.26% | 16 | 26.23% |

**Table 4.4**
(continued)

| Attributes | Count | % | Problem change | | | Problem_Change_Causal | | | | | | | | | |
|---|---|---|---|---|---|---|---|---|---|---|---|---|---|---|---|
| | | | | | | Little or no causal impact | | Modest causal impact | | Balanced causal influence | | Significant causal influence | | Very strong causal influence | |
| Weighted/ unweighted voting | 1 | 0.18% | **worsened considerably** | 0 | 0% | 0 | 0% | 0 | 0% | 0 | 0% | 0 | 0% | 0 | 0% |
| | | | **worsened slightly** | 0 | 0% | 0 | 0% | 0 | 0% | 0 | 0% | 0 | 0% | 0 | 0% |
| | | | **stayed the same** | 1 | 100% | 1 | 100% | 0 | 0% | 0 | 0% | 0 | 0% | 0 | 0% |
| | | | **improved slightly** | 0 | 0% | 0 | 0% | 0 | 0% | 0 | 0% | 0 | 0% | 0 | 0% |
| | | | **improved considerably** | 0 | 0% | 0 | 0% | 0 | 0% | 0 | 0% | 0 | 0% | 0 | 0% |

| | | | | | | | | | | | | | | | |
|---|---|---|---|---|---|---|---|---|---|---|---|---|---|---|---|
| Qualified majority | 57 | 10.25% | worsened considerably | 7 | 12.28% | 1 | 14.29% | 6 | 85.71% | 0 | 0% | 0 | 0% | 0 | 0% |
| | | | worsened slightly | 4 | 7.02% | 2 | 50% | 2 | 50% | 0 | 0% | 0 | 0% | 0 | 0% |
| | | | stayed the same | 5 | 8.77% | 1 | 20% | 0 | 0% | 2 | 40% | 2 | 40% | 0 | 0% |
| | | | improved slightly | 23 | 40.35% | 1 | 4.35% | 7 | 30.43% | 8 | 34.78% | 7 | 30.43% | 0 | 0% |
| | | | improved considerably | 18 | 31.58% | 0 | 0% | 0 | 0% | 10 | 55.56% | 8 | 44.44% | 0 | 0% |

**Table 4.4**
(continued)

| Attributes | Count | % | Problem change | | Problem_Change_Causal | | | | | | | | | |
|---|---|---|---|---|---|---|---|---|---|---|---|---|---|---|
| | | | | | Little or no causal impact | | Modest causal impact | | Balanced causal influence | | Significant causal influence | | Very strong causal influence | |
| Simple majority | 8 | 1.44% | **worsened considerably** | | 0 | 0% | 0 | 0% | 0 | 0% | 0 | 0% | 0 | 0% |
| | | | 0 | 0% | | | | | | | | | | |
| | | | **worsened slightly** | | 0 | 0% | 0 | 0% | 0 | 0% | 0 | 0% | 0 | 0% |
| | | | 0 | 0% | | | | | | | | | | |
| | | | **stayed the same** | | 0 | 0% | 0 | 0% | 0 | 0% | 0 | 0% | 0 | 0% |
| | | | 0 | 0% | | | | | | | | | | |
| | | | **improved slightly** | | 0 | 0% | 2 | 33.33% | 4 | 66.67% | 0 | 0% | 0 | 0% |
| | | | 6 | 75% | | | | | | | | | | |
| | | | **improved considerably** | | 0 | 0% | 0 | 0% | 2 | 100% | 0 | 0% | 0 | 0% |
| | | | 2 | 25% | | | | | | | | | | |

| | | | | | | | | | | | | | | | |
|---|---|---|---|---|---|---|---|---|---|---|---|---|---|---|---|
| Right to opt-out, file objection | 66 | 11.87% | worsened considerably | 9 | 13.64% | 3 | 33.33% | 6 | 66.67% | 0 | 0% | 0 | 0% | 0 | 0% |
| | | | worsened slightly | 25 | 37.88% | 17 | 68% | 2 | 8% | 4 | 16% | 2 | 8% | 0 | 0% |
| | | | stayed the same | 9 | 13.64% | 1 | 11.11% | 0 | 0% | 2 | 22.22% | 6 | 66.67% | 0 | 0% |
| | | | improved slightly | 13 | 19.70% | 3 | 23.08% | 2 | 15.38% | 2 | 15.38% | 6 | 46.15% | 0 | 0% |
| | | | improved considerably | 10 | 15.15% | 0 | 0% | 1 | 10% | 4 | 40% | 5 | 50% | 0 | 0% |
| Total | 556 | 100% | | | | 126 | | 124 | | 94 | | 151 | | 61 | |

and worsened slightly to worsened considerably. Looking only at the top category—improved considerably—we see that these results occur in 31.6 percent of the regimes relying on some form of qualified majority, 20.4 percent of the cases relying on consensus, and 18.42 percent of the arrangements requiring unanimity. A striking finding in this realm is that none of the eleven cases in which there is no explicit decision rule produced a considerable improvement in the problem.

If we combine the two top scores—improved considerably and improved slightly—a somewhat different picture emerges. Regimes relying on some form of qualified majority as a decision rule are associated with problem improvement in 71.9 percent of the cases. The comparable figure for consensual procedures is 49.5 percent of the cases, for unanimity 36.8 percent of the records, and for various systems featuring opt-out procedures 34.6 percent of the cases. Although the association between regimes employing qualified majorities and problem improvement is impressive, it is important to bear in mind that such decision rules are used in only 10.3 percent of the cases included in the IRD. A more general conclusion, then, is that any of a number of decision rules may be in use in regimes associated with problem improvement. But equally important is the observation that problem improvement—even when we combine the top two categories—is a substantially harder hurdle to clear than goal attainment. With the notable exception of the fifty-seven cases featuring some form of qualified majority, problem improvement does not occur in more than half of the cases featuring any of the relevant decision rules.

Here, too, we need to draw attention to the distinction between measures of association between problem improvement and the operation of regimes and the causal role that these arrangements play in bringing about improvement (or for that matter worsening) in the problems at hand. As in the case of goal attainment, we have a five-point ordinal scale regarding causal impact, ranging from little or no impact, through modest impact, balanced impact, and significant impact to very strong impact. Although the number of cases in which a problem improved and the regime made a significant impact varies substantially from one decision rule to another, it is worth emphasizing at the outset that a substantial number of regimes do matter. Perhaps the most important inferences to be drawn from this discussion have to do with consensual procedures. When problems improve considerably and the relevant regimes make use of consensus rules, the causal impact of the regimes is very strong in 26.2 percent of the cases and significant in 44.3 percent of the cases. The parallel figures when problems improve only slightly are 21.8 and 34.5 percent of the cases. The only other decision rule that produces results worthy of comparison with consensus is unanimity. When problems improve considerably, regimes with unanimity rules produce a very strong impact in 28.6 percent of the cases and a significant impact in 23.8 percent of the cases. The parallel figures for cases where problems improve only slightly are 0 and 38.1 percent. One slight anomaly in these data is that regimes operative in situations where the problem worsens considerably and the management system has a significant impact rely on consensual procedures in 22.5 percent of the cases.

What inferences can we draw from these data regarding our theoretical interest in links between decision rules and the effectiveness of regimes? The proportion of cases in which problems improve considerably and regimes make a sizable difference is smaller than the comparable proportion in the case of goal attainment. This is hardly surprising since it is clear that solving problems is generally harder than fulfilling stated goals. That said, however, the evidence suggests that regimes do make a difference in solving problems in a significant number of cases. And in these cases, there is a definite relationship between the decision rules that regimes use and their causal impact on problem solving. Most of the action here centers on regimes that rely on consensual procedures or on unanimity rules. To the extent that consensual procedures loom large in this setting, we can construe this as evidence that is generally supportive of the collective-action perspective on the link between decision rules and institutional effectiveness. But this does not account for the link between effectiveness and the use of unanimity rules. This link suggests that regimes can make a difference in terms of problem solving, even when they remain sensitive to traditional concerns about sovereignty requiring each member of a regime to consent explicitly to decisions made or actions taken in the name of the regime. Overall, there is little support in these inferences for the proposition that relying on consensual procedures and even on unanimity rules activates the law of the least ambitious program. Regimes operating on the basis of consensus or even unanimity can and frequently do play a significant role in alleviating the problems that lead to their creation.

## 4.3 Compliance Mechanisms and Regime Effectiveness

The fact that mechanisms featuring punishments and rewards that are familiar to us in domestic settings are of limited use at the international level does not mean that international environmental regimes have no compliance mechanisms at all. In this section, we inquire first whether regimes have compliance mechanisms and if so, what sorts of mechanisms occur most frequently in this setting. Following a description of current practices in this realm, we turn to an account of theoretical expectations regarding links between compliance mechanisms and effectiveness and to an assessment of data included in the IRD that bear on this issue. As in the case of decision rules, our findings are based on evidence relating to a sizable collection of environmental regimes.

### 4.3.1 Compliance Mechanisms—Current Practice

To explore the links between compliance mechanisms and regime effectiveness, we need to start by clarifying some conceptual matters. Compliance refers to the extent to which the behavior of a regime's subjects conforms to the rules—including both requirements and prohibitions—and meets the programmatic commitments embedded in the provisions of the regime. Note here that the existence or operation of the regime may or may not be the cause of such behavior and that compliance is a matter of concern with respect both to a regime's formal members—usually states—and to actors operating under the jurisdiction of members (e.g., corporations). By contrast, enforcement refers to a particular type of method used to elicit compliant behavior on the part of some identifiable class of actors. In essence, this method seeks to affect behavior by influencing the incentives of individual actors. Many analysts use the term *enforcement* to refer exclusively to the imposition of penalties or the threat of sanctions designed to produce compliance by driving up the costs that actors associate with violations or noncompliant behavior. Yet there is much to be said for broadening this usage to include rewards or

| | Utilitarian reasons | Nonutilitarian reasons |
|---|---|---|
| Unitary actor | Enforcement | Legitimacy/capacity/socialization |
| Nonunitary actor | Management | Legalization/SOPS |

**Figure 4.1**
Approaches to compliance

promises of rewards as a means of strengthening incentives to comply by increasing the benefits associated with compliant behavior. In both cases, enforcement assumes the existence of subjects capable of weighing the benefits and costs associated with specified options, choosing to comply when the expected value of conformance exceeds that of violation, and opting for noncompliance under the opposite conditions. Enforcement undoubtedly constitutes an effective or efficient means of eliciting compliant behavior under some conditions. But it is easy to see that resorting to penalties and rewards does not constitute the only approach to achieving compliance with the provisions of international regimes.

This discussion also suggests a rough but nonetheless useful way to distinguish among major approaches to the achievement of compliance. Figure 4.1 separates the subjects of these provisions into unitary and nonunitary actors and divides the calculations these actors make into utilitarian and nonutilitarian processes. Viewed in this way, the upper-left-hand quadrant of the resulting 2 × 2 matrix constitutes the domain of enforcement. Here, unitary actors engage in benefit-cost calculations and make decisions about whether or not to comply on the basis of these calculations. The upper-right-hand quadrant, by contrast, includes unitary actors whose behavior is driven—at least in part—by factors that are hard to reduce to calculations of benefits and costs. Considerations like the force of habit, the effects of socialization or learning, and the influence of feelings of legitimacy, among others, all belong to this class of motivating forces.

The lower half of the figure replicates the distinction between utilitarian and nonutilitarian sources of behavior but drops the assumption that the relevant actors are unitary decision makers. Thus, individual government agencies or corporations affected by a regime's provisions may base their actions on calculations of benefits and costs to themselves and seek to influence the behavior of the governments of member states accordingly. Similarly, agencies responsible for implementing a regime's provisions may develop standard operating procedures (SOPs) in this realm that produce compliance with requirements and prohibitions in the absence of any conscious deliberation, quite apart from the calculations of the government as a whole. Of course, these distinct sources of (non)compliant behavior are not mutually exclusive. To take a single example, states that are members of a regime may comply with its provisions both because policymakers at the national level expect compliance to produce net benefits and because those agencies responsible for implementation on a day-to-day basis develop SOPs that produce compliant behavior in the absence of case-by-case calculations.

With these distinctions in hand, we can consider the compliance mechanisms in use in international environmental regimes. Most compliance mechanisms associated with environmental regimes belong to the upper half of the table in the sense that they target the governments of states that are the formal members of these institutional arrangements. But determining whether these mechanisms are utilitarian measures that seek to change the incentives of subjects can become a tricky business. The essential character of some mechanisms is evident. Trade sanctions designed to prevent members of the ozone regime from selling ozone-depleting substances to nonmembers are meant to deter potential violators by threatening to impose penalties on those who violate the rule in question. Much the same is true of arrangements, such as those embedded in the Charter of the United Nations, that call for the suspension of the voting rights of members that are more than so far in arrears in the payment of dues or fees. There are as well a number of mechanisms that seek to elicit compliant behavior by promising rewards rather than threatening penalties. The funding arrangements set up under the terms of the Montreal Protocol Multilateral Fund and the Global Environment Facility are cases in point. Much the same can be said of the operation of mechanisms such as Joint Implementation and the Clean Development Mechanism established under the terms of the Kyoto Protocol to the UN Framework Convention on Climate Change.

Other compliance mechanisms directed at the governments of member states, however, rest on premises that are hard to understand in simple utilitarian terms. A common assumption is that many members of international regimes would be willing to comply with the relevant rules and commitments but simply lack the capacity to do so (Chayes and Chayes 1995). What are needed in such cases are mechanisms aimed at capacity building rather than measures featuring penalties or rewards. A number of current practices fall squarely into this category. These include various forms of technology transfer, efforts to train personnel, and procedures designed to make up for the inability of governments in individual member states to engage in the biophysical assessments and behavioral monitoring needed to achieve compliance with the relevant rules and commitments. The introduction of transition periods or grace periods, such as the provision in the ozone-depletion regime allowing developing countries to defer compliance for a period of ten years, can serve as a corollary of these capacity-building measures. What is more, some common compliance mechanisms play roles that are helpful both from a utilitarian perspective and from a nonutilitarian perspective. Consider various reporting requirements and especially systems of implementation review in this connection (Victor, Raustiala, and Skolnikoff 1998). These arrangements can help to drive up the cost of noncompliance by reducing the prospect that violations will go undetected and, as a result, increasing the expected costs of noncompliance. At the same time, systems of implementation review can play important roles in nonutilitarian terms by reassuring individual parties that others are living up to their commitments and, in the process, reinforcing feelings of legitimacy or propriety as a source of compliant behavior.

Although they are less common, some compliance mechanisms associated with environmental regimes belong to the lower half of the figure in the sense that they focus on the actions of subnational or nonstate actors that are not regime members.[4] A distinguishing feature of many environmental regimes is that they target the behavior of subnational actors, including municipal power plants, corporations, and even individuals. Accordingly, compliance mechanisms built into environmental regimes can work by helping member states to elicit compliance from various categories of subnational actors. Often this involves procedures designed to increase the transparency of (non)compliant behavior in states whose governments lack the ability to monitor the behavior of key actors closely. The work of TRAFFIC in conjunction with the regime governing trade in endangered species exemplifies this mechanism. In other cases, nonstate actors assume important roles relating to compliance because they have relevant expertise or experience rather than because governments are lacking in capacity. The role of accreditation societies in certifying that new tankers meet equipment standards set forth in the marine pollution regime (e.g., segregated ballast tanks) is a well-known case in point (Mitchell 1994a). Much the same is true of the role of insurance companies in providing insurance for tankers once they are accredited. Compliance mechanisms can also play a more direct role by allowing nonstate actors to take steps toward dealing with acknowledged problems, whether or not the governments of states are ready and willing to take appropriate action. A striking example arises from the work of the Forest Stewardship Council, which labels wood products that fulfill requirements of sustainable harvesting practices, thereby allowing consumers to act in a way that promotes compliance with the terms of the International Tropical Timber Agreement.

Beyond this, we need to recognize that the rules and commitments of environmental regimes are often ambiguous, a fact that gives rise to disagreements regarding the extent to which specific behavior is (or is not) a violation of the rules and commitments of a regime. Of course, similar concerns arise with regard to regulatory systems at all levels of social organization, and public authorities expend a great deal of time and resources endeavoring to resolve disputes of this sort. What makes international society unusual in this connection is the absence of general arrangements (e.g., a system of courts) capable of resolving such matters in a manner that interested parties will accept as authoritative and legitimate. As a result, it typically falls to individual regimes to establish procedures designed to reconcile disagreements regarding the extent to which the actions of individual parties constitute violations of the relevant rules or commitments. Although many environmental regimes have provisions dealing with dispute settlement, these arrangements seldom become effective as mechanisms for resolving disagreements about (non)-compliance. An interesting alternative, currently emerging under the terms of arrangements such as the ozone-depletion regime, features the establishment and application of noncompliance procedures. The essential premise underlying these procedures is the idea that it is often effective to treat disagreements about compliance as problems to be solved rather than as willful violations whose perpetrators should be punished more or less severely. Although it is too early to pass judgment on the performance of these noncompliance procedures, the fact that they are less adversarial than traditional dispute-settlement procedures may make them better suited to a social setting in which there is no centralized public authority with the capacity to render authoritative judgments about compliance, much less the resources to make such judgments stick.

### 4.3.2 Compliance Mechanisms—Theoretical Expectations

Analysts whose thinking rests on utilitarian premises have a clear point of departure in considering the relationship between compliance mechanisms and the effectiveness of international regimes. They focus squarely on incentives and take the view that regime members will adjust their behavior to fulfill the requirements of these institutional arrangements when the benefits of compliance exceed the costs or, in other words, when opting for compliance produces net benefits. Individual actors may conclude that there are net benefits arising from compliance due to the high costs associated with noncompliance, the magnitude of the expected benefits arising from compliance, or some combination of the two. But the usual assumption is that compliance is costly, a fact that tends to place the burden of proof on those who claim that choosing to comply will yield net benefits.

There is a pronounced tendency among those who subscribe to this perspective to assume that the key to achieving high levels of compliance lies in the role of enforcement, and more specifically in the imposition of penalties or the threat of penalties to drive up the costs of noncompliance and, as a result, to deter violations. It is this way of thinking about the problem of compliance coupled with the underdeveloped character of mechanisms at the international level capable of imposing sanctions that leads many observers to dismiss the idea that regimes can be effective sources of governance and, more generally, to raise questions about the extent to which international law deserves to be treated as law in any normal sense of the term. What is more, those who adhere to this line of thinking are apt to argue that achieving compliance with the rules and commitments of international regimes will become increasingly difficult as the depth of cooperation and, as a result, the extent of the behavioral requirements of compliance increase (Downs, Rocke, and Barsoom 1996; Barrett 2003). As the costs of compliance rise, in other words, the importance of enforcement as a means of driving up the costs of non-compliance and ensuring that compliance produces net benefits grows.

How persuasive is this argument? Pointing to the occurrence of relatively high levels of compliance with the rules and commitments of specific regimes, a sizable group of analysts have advanced the view that penalties or threats of penalties are not the critical determinant of compliance at the international level. Members of this group advance a number of related propositions. They suggest, to begin with, that rewards in such forms as technology transfers and financial assistance may be sufficient to give many regime members incentives to comply with a range of rules and commitments. These analysts remain generally committed to a utilitarian mode of reasoning. But their take on the sources of compliance is not limited to pointing out the relevance of rewards. They argue, for instance, that mechanisms that increase transparency regarding the behavior of regime members can make a difference, even if they do not lead to the imposition of penalties as such. The idea here is that many actors who are prone to violate rules when they can do so privately or clandestinely will comply with rules and commitments when they sense that it is highly likely that their behavior will come to light as a matter of public knowledge. Many of these observers argue as well that efforts to elicit compliance are better thought of in terms of the concept of management than in terms of enforcement (Chayes and Chayes 1995). They take the view that compliance mechanisms featuring the sharing of information, the provision of technical assistance, the mobilization of financial aid, and the establishment of systems of implementation review will be just as effective in producing compliance as conventional enforcement procedures.

All these arguments share the assumption that regime members routinely make calculations of benefits and costs and comply when they anticipate that compliant behavior will produce net benefits. In short, they all lead to the expectation that there should be a discernible correlation between the operation of well-defined compliance mechanisms and the effectiveness of international regimes. Yet many of those interested in compliance—at all levels of social organization—have observed that any institutional arrangement dependent solely on the operation of compliance mechanisms rooted in utilitarian thinking to elicit behavioral conformance is bound to fail. While the imposition of penalties or the provision of rewards may prove effective at the margin, even well-endowed public authorities would run into trouble right away unless most subjects complied with the relevant rules and commitments most of the time without regard to the impacts of punishments and rewards.

How is this possible? This is where arguments rooted in a social-practice conception of institutions come into focus. On this account, actors are likely to comply with the provisions of regimes for reasons that do not lend themselves to calculations of benefits and costs and that do not treat noncompliance as a default option that must be overcome through the establishment of mechanisms capable of altering the incentives of individual actors. To the extent that regimes create social practices and assign roles to participating actors, individual subjects may not even consider the option of violating the rules of the game associated with specific practices. Much of the behavior of most regime members is routinized and soon becomes a matter of second nature. Once a pattern of compliant behavior is established and is incorporated into the standard operating procedures of administrative agencies within individual regime members, compliant behavior may become habitual in the sense that it does not trigger deliberation on a case-by-case basis. Equally important is the fact that regime members who view an institutional arrangement as legitimate may simply regard compliance as the right and proper thing to do. They are motivated, in effect, more by the logic of appropriateness than by utilitarian calculations in which compliant behavior reflects the presence and operation of mechanisms capable of shifting benefits and costs in such a way as to provide actors with incentives to comply. Compliance, in this account, is more a matter of the overall character of a regime and the attitudes of regime members toward it than a matter of the operation of well-defined mechanisms that lead members to conclude that compliant behavior will yield net benefits.

The expectations associated with these arguments regarding what we will find in examining the empirical evidence are straightforward. Those who think that enforcement is the key to compliance will expect to find a high correlation between the operation of mechanisms capable of generating rewards and especially penalties and the success of the relevant regimes. Those who subscribe to what is often called the management perspective on compliance will expect to find a clear relationship between the effectiveness of regimes and the operation of compliance mechanisms, but not necessarily mechanisms that work through the use of penalties and rewards. Still others whose thinking is rooted more in a social-practice perspective on regimes will expect to find no clearcut relationship between institutional success and the operation of specific compliance mechanisms. Of course, they do not take the view that the presence of compliance mechanisms is detrimental to the achievement of success. Rather, they simply do not expect to find any well-defined relationship between the two.

### 4.3.3 Compliance Mechanisms—Empirical Evidence

As in the case of our analysis of links between decision rules and effectiveness in the previous section, we turn now to an examination of evidence contained in the IRD that can shed light on the links between compliance mechanisms and regime effectiveness. Part II of the IRD includes a section on compliance mechanisms encompassing a number of specific mechanisms that can stand as independent variables in our efforts to sort out the merits of various theoretical arguments regarding the roles that these mechanisms play as determinants of regime effectiveness. Here we follow a procedure similar to the one employed in our analysis of decision rules. Specifically, we focus on a few key variables in the interests of maintaining tractability and deriving some initial results regarding these links that are clear and easy to understand.

The variables goal attainment and problem change constitute the dependent variables in this subsection, just as they did in the analysis of the links between decision rules and effectiveness. But now we have new independent variables. To begin with, we turn to the question in the IRD that asks: "What formal compliance mechanisms are provided for in the regime's constitutive provisions to achieve compliance?" In answering this question, coders were allowed to choose among nine specific responses starting with no compliance mechanisms and ranging through a variety of mechanisms involving membership rights and penalties to an option called dissolution of linkages and meant to refer to the fragmentation or implosion of the regime. This variable recognizes that regimes often provide—at least formally—for a range of compliance mechanisms and seeks to establish whether there are discernible patterns linking the presence of particular mechanisms to goal attainment or problem change.

This examination of the role of compliance mechanisms also seeks to address the debate about what has become known in recent literature as the distinction between enforcement and management approaches to compliance (Chayes and Chayes 1995). Thus, the IRD asks: "Do these [compliance] procedures reflect an enforcement approach or a management approach to compliance?" Naturally, there are actual cases in which regimes make use of some combination of these procedures. Nevertheless, the IRD contains an extensive set of responses to this question, a fact that makes it possible to pose inquiries about the primary approaches to compliance used and about results in terms of goal attainment and problem change.

Once again, we are able to construct select queries in MS Access to probe the evidence relating to these links. As in the case of decision rules, we have developed a set of four specific queries:

*Query 5* Are their clearcut links between the nature of the compliance mechanisms provided and goal attainment?

*Query 6* Are there discernible patterns in the connections between the nature of the compliance mechanisms provided and problem change?

*Query 7* Are there connections between the approach to compliance chosen and goal attainment?

*Query 8* Are there links between the approach to compliance chosen and problem change?

What can we learn from these queries, and what light can they shed on our underlying concern regarding the effectiveness of international environmental regimes? To answer these questions, we start with the first pair of queries, leaving an analysis of the remaining queries to a later part of this subsection. These queries direct attention to the roles that specific compliance mechanisms play in helping regimes to attain their stated goals and to contribute to solving the problems that led to their creation. The data these queries generate are reported in tables 4.5 and 4.6.

An examination of table 4.5 reveals that only a few of the members of the potential set of compliance mechanisms figure prominently in efforts to fulfill the stated goals of regimes. Regimes use capacity building aimed at enhancing compliance in 25.9 percent of the cases. The issuance of notices of violation occurs in 24 percent of the cases. Arrangements intended to improve compliance by allowing for transitional or grace periods figure in 21.5 percent of the cases. The only other compliance mechanisms that come into play in more than 5 percent of the cases are efforts to identify case-specific procedures (5.3 percent of the cases) and the use of financial sanctions (5.7 percent of the cases).

This means that we can confine our attention to a relatively narrow range of relationships in asking about the links between compliance mechanisms and both the attainment of goals and the causal significance of regimes in producing such results. The case of capacity building, for instance, yields an encouraging story with regard to the effectiveness of environmental regimes. Thus, goal attainment occurred in 89.2 percent of the cases featuring capacity building, and—most interestingly—the coders judged the regime to have had a large causal impact in 83.6 percent of the cases in which governance systems involving capacity building succeeded in attaining goals. Somewhat similar results emerge from an examination of cases featuring the use of notices of violation. Of the cases included in this category, 90.8 percent led to the attainment of goals, and the coders concluded that the regime had a large causal influence in 81.2 percent of the cases leading to the attainment of goals.

The results of granting transitional periods to encourage compliance, on the other hand, are less striking. Although goal attainment occurred in 86.8 percent of these cases, the regime had a large causal influence in only 35.6 percent of the cases in this category (a finding moderated somewhat by the fact that regimes produced a modest causal influence in 62.7 percent of the cases). Though less striking, a few other inferences to be drawn from the data presented in table 4.5 are worth noting in passing. The absence of formal compliance mechanisms does not necessarily mean that regimes cannot contribute to goal attainment. The number of cases is small, but the imposition of financial or economic penalties can lead to goal attainment through processes in which regimes have a large causal influence. As might be expected, moreover, cases in which compliance mechanisms are customized to the problem at hand yield positive results through processes in which regimes make a difference.

The links between compliance mechanisms and problem solving are broadly similar, though significantly more complex. Query 6 produced one rather striking effect that did not stand out in Query 5. Thus, the coders found no compliance mechanism provided in 29.7 percent of the cases in this query. In our judgment, this is not a peculiar result. In fact, the numbers reported in table 4.6 may well offer a more accurate representation of the conditions prevailing in the universe of cases than the parallel numbers in table 4.5.[5] Apart from this somewhat anomalous difference, the pattern of compliance mechanisms in Query 6 is broadly comparable to the picture arising from Query 5. Thus, 19.1 percent of the regimes in this set make use of notices of violation; capacity-building measures are in use in 16.7 percent of the cases, and 8.9 percent of the cases feature arrangements involving transitional periods. Interestingly, 4.1 percent of the cases included in this query make provisions for the suspension of membership rights as a compliance mechanism, a finding indicating that interest in more coercive approaches to compliance may be on the rise.

What about the links between the use of various compliance mechanisms and progress toward problem solving? This query indicates that the presence of explicit compliance mechanisms is not a necessary condition for progress. In 20.6 percent of the cases lacking compliance mechanisms, there was considerable improvement in the problem; slight improvement occurred in another 24.7 percent of the cases. Among those cases featuring considerable improvement in the problem, the regime had a very strong causal influence in 13.3 percent of the cases and a significant causal influence in another 46.7 percent of the cases. This certainly does not license the conclusion that compliance mechanisms are unimportant in the search for effectiveness. But it does lend some support to the social-practice argument that engaging actors in more or less complex social networks can produce positive results, quite apart from the development of compliance mechanisms (Reinecke 1998).

**Table 4.5**
Compliance mechanisms and goal attainment

| Compliance | Count | % | Goals fulfilled | | | Degree of Causality | | | | | | | |
|---|---|---|---|---|---|---|---|---|---|---|---|---|---|
| | | | | | | Little or no causal impact | | Modest causal influence | | Large causal influence | | Negative causal influence | |
| No compliance mechanisms | 11 | 3.47% | No | 6 | 54.55% | 3 | 50% | 3 | 50% | 0 | 0% | 0 | 0% |
| | | | Yes | 5 | 45.45% | 0 | 0% | 2 | 40% | 3 | 60% | 0 | 0% |
| Compliance Committee to review compliance by members with conservation rules | 5 | 1.58% | No | 0 | 0% | 0 | 0% | 0 | 0% | 0 | 0% | 0 | 0% |
| | | | Yes | 5 | 100% | 0 | 0% | 0 | 0% | 5 | 100% | 0 | 0% |
| Additional compliance mechanisms not covered by data protocol | 17 | 5.36% | No | 0 | 0% | 0 | 0% | 0 | 0% | 0 | 0% | 0 | 0% |
| | | | Yes | 17 | 100% | 3 | 17.65% | 0 | 0% | 14 | 82.35% | 0 | 0% |

| | | | | | | | | | | | |
|---|---|---|---|---|---|---|---|---|---|---|---|
| Observer of a violation can notify commission of same; commission does not itself initiate such notices | 5 | 1.58% | No: 2 (40%) | 0 | 0% | 2 | 100% | 0 | 0% | 0 | 0% |
| | | | Yes: 3 (60%) | 0 | 0% | 2 | 66.67% | 1 | 33.33% | 0 | 0% |
| Provision for mutual monitoring of nationals engaged in fishing | 5 | 1.58% | No: 2 (40%) | 0 | 0% | 2 | 100% | 0 | 0% | 0 | 0% |
| | | | Yes: 3 (60%) | 0 | 0% | 2 | 66.67% | 1 | 33.33% | 0 | 0% |
| Issuance of notices of violations | 76 | 23.97% | No: 7 (9.21%) | 3 | 42.86% | 3 | 42.86% | 0 | 0% | 1 | 14.29% |
| | | | Yes: 69 (90.79%) | 4 | 5.80% | 9 | 13.04% | 56 | 81.16% | 0 | 0% |
| Issuance of notices of violations (EU only) | 6 | 1.89% | No: 0 (0%) | 0 | 0% | 0 | 0% | 0 | 0% | 0 | 0% |
| | | | Yes: 6 (100%) | 0 | 0% | 1 | 17% | 5 | 83% | 0 | 0% |
| Suspension of membership rights | 5 | 1.58% | No: 1 (20%) | 0 | 0% | 1 | 100% | 0 | 0% | 0 | 0% |
| | | | Yes: 4 (80%) | 0 | 0% | 2 | 50% | 2 | 50% | 0 | 0% |

**Table 4.5**
(continued)

| Compliance | Count | % | Goals fulfilled | | Degree of Causality | | | | | | | |
|---|---|---|---|---|---|---|---|---|---|---|---|---|
| | | | | | Little or no causal impact | | Modest causal influence | | Large causal influence | | Negative causal influence | |
| Exclusion from membership | 5 | 1.58% | No | | 0 | 0% | 0 | 0% | 0 | 0% | 0 | 0% |
| | | | 0 | 0% | | | | | | | | |
| | | | Yes | | 0 | 0% | 0 | 0% | 5 | 100% | 0 | 0% |
| | | | 5 | 100% | | | | | | | | |
| Imposition of military punishments | 5 | 1.58% | No | | 0 | 0% | 0 | 0% | 0 | 0% | 0 | 0% |
| | | | 0 | 0% | | | | | | | | |
| | | | Yes | | 0 | 0% | 0 | 0% | 5 | 100% | 0 | 0% |
| | | | 5 | 100% | | | | | | | | |
| Imposition of financial/economic punishments | 18 | 5.68% | No | | 0 | 0% | 0 | 0% | 0 | 0% | 0 | 0% |
| | | | 0 | 0% | | | | | | | | |
| | | | Yes | | 1 | 5.56% | 2 | 11.11% | 15 | 83.33% | 0 | 0% |
| | | | 18 | 100% | | | | | | | | |

| | | | | | | | | | | | | | |
|---|---|---|---|---|---|---|---|---|---|---|---|---|---|
| Imposition of financial/economic punishments (EU only) | 6 | 1.89% | No | 0 | 0% | 0 | 0% | 0 | 0% | 0 | 0% | 0 | 0% |
| | | | Yes | 6 | 100% | 0 | 0% | 1 | 16.67% | 5 | 83.33% | 0 | 0% |
| Support for capacity building to enhance compliance | 82 | 25.87% | No | 9 | 10.98% | 3 | 33.33% | 5 | 55.56% | 1 | 11.11% | 0 | 0% |
| | | | Yes | 73 | 89.02% | 3 | 4.11% | 9 | 12.33% | 61 | 83.56% | 0 | 0% |
| Granting of a transition period to achieve compliance | 68 | 21.45% | No | 9 | 13.24% | 3 | 33.33% | 5 | 55.56% | 1 | 11.11% | 0 | 0% |
| | | | Yes | 59 | 86.76% | 1 | 1.69% | 37 | 62.71% | 21 | 35.59% | 0 | 0% |
| Compliance Review | 3 | 0.95% | No | 0 | 0% | 0 | 0% | 0 | 0% | 0 | 0% | 0 | 0% |
| | | | Yes | 3 | 100% | 0 | 0% | 1 | 33.33% | 2 | 66.67% | 0 | 0% |
| Total | 317 | 100.00% | | | | 24 | | 89 | | 203 | | 1 | |

**Table 4.6**
Compliance mechanisms and problem change

| Attributes | Count | % | Problem Change | | Problem_Change_Causal | | | | | | | | | |
|---|---|---|---|---|---|---|---|---|---|---|---|---|---|---|
| | | | | | Little or no causal impact | | Modest causal impact | | Balanced causal influence | | Significant causal influence | | Very strong causal influence | |
| No compliance mechanisms | 73 | 29.67% | **worsened considerably** | | 6 | 60% | 4 | 40% | 0 | 0% | 0 | 0% | 0 | 0% |
| | | | 10 | 13.70% | | | | | | | | | | |
| | | | **worsened slightly** | | 12 | 70.59% | 5 | 29.41% | 0 | 0% | 0 | 0% | 0 | 0% |
| | | | 17 | 23.29% | | | | | | | | | | |
| | | | **stayed the same** | | 2 | 15.38% | 1 | 7.69% | 1 | 7.69% | 7 | 53.85% | 2 | 15.38% |
| | | | 13 | 17.81% | | | | | | | | | | |
| | | | **improved slightly** | | 5 | 27.78% | 4 | 22.22% | 8 | 44.44% | 1 | 5.56% | 0 | 0% |
| | | | 18 | 24.66% | | | | | | | | | | |
| | | | **improved considerably** | | 0 | 0% | 2 | 13.33% | 4 | 26.67% | 7 | 46.67% | 2 | 13.33% |
| | | | 15 | 20.55% | | | | | | | | | | |

| | | | | | | | | | | | | | | | |
|---|---|---|---|---|---|---|---|---|---|---|---|---|---|---|---|
| Compliance Committee to review compliance by members with conservation rules | 1 | 0.41% | **worsened considerably** | 0 | 0% | 0 | 0% | 0 | 0% | 0 | 0% | 0 | 0% | 0 | 0% |
| | | | **worsened slightly** | 0 | 0% | 0 | 0% | 0 | 0% | 0 | 0% | 0 | 0% | 0 | 0% |
| | | | **stayed the same** | 0 | 0% | 0 | 0% | 0 | 0% | 0 | 0% | 0 | 0% | 0 | 0% |
| | | | **improved slightly** | 1 | 100% | 0 | 0% | 0 | 0% | 0 | 0% | 1 | 100% | 0 | 0% |
| | | | **improved considerably** | 0 | 0% | 0 | 0% | 0 | 0% | 0 | 0% | 0 | 0% | 0 | 0% |

**Table 4.6**
(continued)

| Attributes | Count | % | Problem Change | | | Problem_Change_Causal | | | | | | | | | |
|---|---|---|---|---|---|---|---|---|---|---|---|---|---|---|---|
| | | | | | | Little or no causal impact | | Modest causal impact | | Balanced causal influence | | Significant causal influence | | Very strong causal influence | |
| Dispute settlement clause in Article XI | 15 | 6.10% | **worsened considerably** | 1 | 6.67% | 1 | 100% | 0 | 0% | 0 | 0% | 0 | 0% | 0 | 0% |
| | | | **worsened slightly** | 6 | 40.00% | 3 | 50% | 0 | 0% | 2 | 33.33% | 1 | 16.67% | 0 | 0% |
| | | | **stayed the same** | 5 | 33% | 0 | 0% | 0 | 0% | 1 | 20% | 4 | 80% | 0 | 0% |
| | | | **improved slightly** | 2 | 13.33% | 0 | 0% | 0 | 0% | 0 | 0% | 2 | 100% | 0 | 0% |
| | | | **improved considerably** | 1 | 6.67% | 0 | 0% | 1 | 100% | 0 | 0% | 0 | 0% | 0 | 0% |

| | | | | | | | | | | | | | | |
|---|---|---|---|---|---|---|---|---|---|---|---|---|---|---|
| Identify additional compliance mechanism, if applicable | 4 | 1.63% | **worsened considerably** | | 0 | 0% | 0 | 0% | 0 | 0% | 0 | 0% | 0 | 0% |
| | | | 0 | 0% | | | | | | | | | | |
| | | | **worsened slightly** | | 0 | 0% | 0 | 0% | 0 | 0% | 0 | 0% | 0 | 0% |
| | | | 0 | 0% | | | | | | | | | | |
| | | | **stayed the same** | | 0 | 0% | 4 | 100% | 0 | 0% | 0 | 0% | 0 | 0% |
| | | | 4 | 100% | | | | | | | | | | |
| | | | **improved slightly** | | 0 | 0% | 0 | 0% | 0 | 0% | 0 | 0% | 0 | 0% |
| | | | 0 | 0% | | | | | | | | | | |
| | | | **improved considerably** | | 0 | 0% | 0 | 0% | 0 | 0% | 0 | 0% | 0 | 0% |
| | | | 0 | 0% | | | | | | | | | | |

**Table 4.6**
(continued)

| Attributes | Count | % | Problem Change | | Problem_Change_Causal: Little or no causal impact | | Modest causal impact | | Balanced causal influence | | Significant causal influence | | Very strong causal influence | |
|---|---|---|---|---|---|---|---|---|---|---|---|---|---|---|
| Observer of a violation can notify commission of same; commission does not itself initiate such notices | 6 | 2.44% | **worsened considerably** | | 1 | 100% | 0 | 0% | 0 | 0% | 0 | 0% | 0 | 0% |
| | | | 1 | 16.67% | | | | | | | | | | |
| | | | **worsened slightly** | | 1 | 25% | 0 | 0% | 2 | 50% | 1 | 25% | 0 | 0% |
| | | | 4 | 66.67% | | | | | | | | | | |
| | | | **stayed the same** | | 0 | 0% | 0 | 0% | 1 | 100% | 0 | 0% | 0 | 0% |
| | | | 1 | 16.67% | | | | | | | | | | |
| | | | **improved slightly** | | 0 | 0% | 0 | 0% | 0 | 0% | 0 | 0% | 0 | 0% |
| | | | 0 | 0% | | | | | | | | | | |
| | | | **improved considerably** | | 0 | 0% | 0 | 0% | 0 | 0% | 0 | 0% | 0 | 0% |
| | | | 0 | 0% | | | | | | | | | | |

| | | | | | | | | | | | | | | | |
|---|---|---|---|---|---|---|---|---|---|---|---|---|---|---|---|
| Provision for mutual monitoring of nationals engaged in fishing | 6 | 2.44% | worsened considerably | 1 | 16.67% | 1 | 100% | 0 | 0% | 0 | 0% | 0 | 0% | 0 | 0% |
| | | | worsened slightly | 4 | 66.67% | 1 | 25% | 0 | 0% | 2 | 50% | 1 | 25% | 0 | 0% |
| | | | stayed the same | 1 | 16.67% | 0 | 0% | 0 | 0% | 1 | 100% | 0 | 0% | 0 | 0% |
| | | | improved slightly | 0 | 0% | 0 | 0% | 0 | 0% | 0 | 0% | 0 | 0% | 0 | 0% |
| | | | improved considerably | 0 | 0% | 0 | 0% | 0 | 0% | 0 | 0% | 0 | 0% | 0 | 0% |

**Table 4.6**
(continued)

| Attributes | Count | % | Problem Change | | Problem_Change_Causal | | | | | | | | | |
|---|---|---|---|---|---|---|---|---|---|---|---|---|---|---|
| | | | | | Little or no causal impact | | Modest causal impact | | Balanced causal influence | | Significant causal influence | | Very strong causal influence | |
| Don't know | 4 | 1.63% | **worsened considerably** | | 1 | 100% | 0 | 0% | 0 | 0% | 0 | 0% | 0 | 0% |
| | | | 1 | 25% | | | | | | | | | | |
| | | | **worsened slightly** | | 0 | 0% | 2 | 100% | 0 | 0% | 0 | 0% | 0 | 0% |
| | | | 2 | 50% | | | | | | | | | | |
| | | | **stayed the same** | | 0 | 0% | 0 | 0% | 1 | 100% | 0 | 0% | 0 | 0% |
| | | | 0 | 0% | | | | | | | | | | |
| | | | **improved slightly** | | 0 | 0% | 0 | 0% | 0 | 0% | 0 | 0% | 0 | 0% |
| | | | 0 | 0% | | | | | | | | | | |
| | | | **improved considerably** | | 0 | 0% | 1 | 100% | 0 | 0% | 0 | 0% | 0 | 0% |
| | | | 1 | 25% | | | | | | | | | | |

| | | | | | | | | | | | | | |
|---|---|---|---|---|---|---|---|---|---|---|---|---|---|
| Issuance of notices of violations | 47 | 19.11% | worsened considerably | | 0 | 0% | 3 | 60% | 0 | 0% | 1 | 20% | 1 | 20% |
| | | | 5 | 10.64% | | | | | | | | | | |
| | | | worsened slightly | | 5 | 83.33% | 0 | 0% | 0 | 0% | 1 | 16.67% | 0 | 0% |
| | | | 6 | 12.77% | | | | | | | | | | |
| | | | stayed the same | | 1 | 11.11% | 5 | 55.56% | 0 | 0% | 1 | 11.11% | 2 | 22.22% |
| | | | 9 | 19.15% | | | | | | | | | | |
| | | | improved slightly | | 1 | 6.67% | 2 | 13.33% | 1 | 6.67% | 2 | 13.33% | 9 | 60% |
| | | | 15 | 31.91% | | | | | | | | | | |
| | | | improved considerably | | 1 | 8.33% | 1 | 8.33% | 2 | 16.67% | 5 | 41.67% | 3 | 25% |
| | | | 12 | 25.53% | | | | | | | | | | |

**Table 4.6**
(continued)

| Attributes | Count | % | Problem Change | | | Problem_Change_Causal | | | | | | | | | |
|---|---|---|---|---|---|---|---|---|---|---|---|---|---|---|---|
| | | | | | | Little or no causal impact | | Modest causal impact | | Balanced causal influence | | Significant causal influence | | Very strong causal influence | |
| Issuance of notices of violations (EU only) | 2 | 0.81% | worsened considerably | 0 | 0% | 0 | 0% | 0 | 0% | 0 | 0% | 0 | 0% | 0 | 0% |
| | | | worsened slightly | 1 | 50% | 0 | 0% | 1 | 100% | 0 | 0% | 0 | 0% | 0 | 0% |
| | | | stayed the same | 0 | 0% | 0 | 0% | 0 | 0% | 0 | 0% | 0 | 0% | 0 | 0% |
| | | | improved slightly | 1 | 50% | 0 | 0% | 0 | 0% | 0 | 0% | 0 | 0% | 0 | 0% |
| | | | improved considerably | 0 | 0% | 0 | 0% | 0 | 0% | 0 | 0% | 1 | 100% | 0 | 0% |

| | | | | | | | | | | | | | | |
|---|---|---|---|---|---|---|---|---|---|---|---|---|---|---|
| Suspension of membership rights | 10 | 4.07% | **worsened considerably** | | **0** | 0.00% | **2** | 66.67% | **0** | 0% | **1** | 33.33% | **0** | 0% |
| | | | 3 | 30% | | | | | | | | | | |
| | | | **worsened slightly** | | **0** | 0% | **0** | 0% | **0** | 0% | **0** | 0% | **0** | 0% |
| | | | 0 | 0% | | | | | | | | | | |
| | | | **stayed the same** | | **1** | 20% | **4** | 80% | **0** | 0% | **0** | 0% | **0** | 0% |
| | | | 5 | 50% | | | | | | | | | | |
| | | | **improved slightly** | | **0** | 0% | **1** | 100% | **0** | 0% | **0** | 0% | **0** | 0% |
| | | | 1 | 10% | | | | | | | | | | |
| | | | **improved considerably** | | **0** | 0% | **0** | 0% | **1** | 100% | **0** | 0% | **0** | 0% |
| | | | 1 | 10% | | | | | | | | | | |

**Table 4.6**
(continued)

| Attributes | Count | % | Problem Change | | Problem_Change_Causal | | | | | | | | | |
|---|---|---|---|---|---|---|---|---|---|---|---|---|---|---|
| | | | | | Little or no causal impact | | Modest causal impact | | Balanced causal influence | | Significant causal influence | | Very strong causal influence | |
| Exclusion from membership | 3 | 1.22% | worsened considerably | | 0 | 0% | 0 | 0% | 0 | 0% | 0 | 0% | 0 | 0% |
| | | | 0 | 0% | | | | | | | | | | |
| | | | worsened slightly | | 0 | 0% | 0 | 0% | 0 | 0% | 0 | 0% | 0 | 0% |
| | | | 0 | 0% | | | | | | | | | | |
| | | | stayed the same | | 1 | 100% | 0 | 0% | 0 | 0% | 0 | 0% | 0 | 0% |
| | | | 1 | 33.33% | | | | | | | | | | |
| | | | improved slightly | | 0 | 0% | 1 | 100% | 0 | 0% | 0 | 0% | 0 | 0% |
| | | | 1 | 33.33% | | | | | | | | | | |
| | | | improved considerably | | 0 | 0% | 0 | 0% | 0 | 0% | 1 | 100% | 0 | 0% |
| | | | 1 | 33.33% | | | | | | | | | | |

| | | | | | | | | | | | | | | | |
|---|---|---|---|---|---|---|---|---|---|---|---|---|---|---|---|
| Imposition of military punishments | 1 | 0.41% | worsened considerably | 0 | 0% | 0 | 0% | 0 | 0% | 0 | 0% | 0 | 0% | 0 | 0% |
| | | | worsened slightly | 0 | 0% | 0 | 0% | 0 | 0% | 0 | 0% | 0 | 0% | 0 | 0% |
| | | | stayed the same | 0 | 0% | 0 | 0% | 0 | 0% | 0 | 0% | 0 | 0% | 0 | 0% |
| | | | improved slightly | 0 | 0% | 0 | 0% | 0 | 0% | 0 | 0% | 0 | 0% | 0 | 0% |
| | | | improved considerably | 1 | 100% | 0 | 0% | 0 | 0% | 0 | 0% | 1 | 100% | 0 | 0% |

**Table 4.6**
(continued)

| Attributes | Count | % | Problem Change | | | Problem_Change_Causal | | | | | | | | | |
|---|---|---|---|---|---|---|---|---|---|---|---|---|---|---|---|
| | | | | | | Little or no causal impact | | Modest causal impact | | Balanced causal influence | | Significant causal influence | | Very strong causal influence | |
| Imposition of financial/economic punishments | 5 | 2.03% | worsened considerably | 0 | 0% | 0 | 0% | 0 | 0% | 0 | 0% | 0 | 0% | 0 | 0% |
| | | | worsened slightly | 0 | 0% | 0 | 0% | 0 | 0% | 0 | 0% | 0 | 0% | 0 | 0% |
| | | | stayed the same | 0 | 0% | 0 | 0% | 0 | 0% | 0 | 0% | 0 | 0% | 0 | 0% |
| | | | improved slightly | 1 | 20% | 0 | 0% | 0 | 0% | 0 | 0% | 1 | 100% | 0 | 0% |
| | | | improved considerably | 4 | 80% | 0 | 0% | 0 | 0% | 0 | 0% | 3 | 75% | 1 | 25% |

| Imposition of financial/economic punishments (EU only) | 2 | 0.81% | worsened considerably | 0 | 0% | 0 | 0% | 0 | 0% | 0 | 0% | 0 | 0% | 0 | 0% |
|---|---|---|---|---|---|---|---|---|---|---|---|---|---|---|---|
| | | | worsened slightly | 1 | 50% | 0 | 0% | 1 | 100% | 0 | 0% | 0 | 0% | 0 | 0% |
| | | | stayed the same | 0 | 0% | 0 | 0% | 0 | 0% | 0 | 0% | 0 | 0% | 0 | 0% |
| | | | improved slightly | 1 | 50% | 0 | 0% | 0 | 0% | 0 | 0% | 1 | 100% | 0 | 0% |
| | | | improved considerably | 0 | 0% | 0 | 0% | 0 | 0% | 0 | 0% | 0 | 0% | 0 | 0% |

**Table 4.6**
(continued)

| Attributes | Count | % | Problem Change | | Problem_Change_Causal | | | | | | | | | |
|---|---|---|---|---|---|---|---|---|---|---|---|---|---|---|
| | | | | | Little or no causal impact | | Modest causal impact | | Balanced causal influence | | Significant causal influence | | Very strong causal influence | |
| Support for capacity building to enhance compliance | 41 | 16.67% | **worsened considerably** | | 0 | 0% | 0 | 0% | 0 | 0% | 1 | 50% | 1 | 50% |
| | | | 2 | 4.88% | | | | | | | | | | |
| | | | **worsened slightly** | | 5 | 100% | 0 | 0% | 0 | 0% | 0 | 0% | 0 | 0% |
| | | | 5 | 12.20% | | | | | | | | | | |
| | | | **stayed the same** | | 1 | 14.29% | 5 | 71.43% | 0 | 0% | 1 | 14.29% | 0 | 0% |
| | | | 7 | 17.07% | | | | | | | | | | |
| | | | **improved slightly** | | 1 | 5.56% | 2 | 11.11% | 1 | 5.56% | 6 | 33.33% | 8 | 44.44% |
| | | | 18 | 43.90% | | | | | | | | | | |
| | | | **improved considerably** | | 0 | 0% | 0 | 0% | 1 | 11.11% | 5 | 55.56% | 3 | 33.33% |
| | | | 9 | 21.95% | | | | | | | | | | |

| | | | | | | | | | | | | | | | |
|---|---|---|---|---|---|---|---|---|---|---|---|---|---|---|---|
| Granting of a transition period to achieve compliance | 22 | 8.94% | worsened considerably | 1 | 4.55% | 0 | 0% | 0 | 0% | 0 | 0% | 1 | 100% | 0 | 0% |
| | | | worsened slightly | 3 | 13.64% | 0 | 0% | 3 | 100% | 0 | 0% | 0 | 0% | 0 | 0% |
| | | | stayed the same | 5 | 22.73% | 1 | 20% | 4 | 80% | 0 | 0% | 0 | 0% | 0 | 0% |
| | | | improved slightly | 4 | 18.18% | 1 | 25% | 2 | 50% | 1 | 25% | 0 | 0% | 0 | 0% |
| | | | improved considerably | 9 | 40.91% | 1 | 11.11% | 0 | 0.00% | 0 | 0% | 6 | 66.67% | 2 | 22.22% |

**Table 4.6**
(continued)

| Attributes | Count | % | Problem Change | | Problem_Change_Causal | | | | | | | | | |
|---|---|---|---|---|---|---|---|---|---|---|---|---|---|---|
| | | | | | Little or no causal impact | | Modest causal impact | | Balanced causal influence | | Significant causal influence | | Very strong causal influence | |
| Compliance Review | 3 | 1.22% | worsened considerably | | 0 | 0% | 0 | 0% | 0 | 0% | 0 | 0% | 0 | 0% |
| | | | 0 | 0% | | | | | | | | | | |
| | | | worsened slightly | | 0 | 0% | 0 | 0% | 0 | 0% | 0 | 0% | 0 | 0% |
| | | | 0 | 0% | | | | | | | | | | |
| | | | stayed the same | | 0 | 0% | 0 | 0% | 0 | 0% | 0 | 0% | 0 | 0% |
| | | | 0 | 0% | | | | | | | | | | |
| | | | improved slightly | | 0 | 0% | 0 | 0% | 2 | 100% | 0 | 0% | 0 | 0% |
| | | | 2 | 66.67% | | | | | | | | | | |
| | | | improved considerably | | 0 | 0% | 0 | 0% | 1 | 100% | 0 | 0% | 0 | 0% |
| | | | 1 | 33.33% | | | | | | | | | | |

| | | | | | | | | | | | | | | | |
|---|---|---|---|---|---|---|---|---|---|---|---|---|---|---|---|
| There are several mechanisms to review compliance, but not of the nature specified in the coding schedule. No punitive measures. | 1 | 0.41% | **worsened considerably** | 0 | 0% | 0 | 0% | 0 | 0% | 0 | 0% | 0 | 0% | 0 | 0% |
| | | | **worsened slightly** | 0 | 0% | 0 | 0% | 0 | 0% | 0 | 0% | 0 | 0% | 0 | 0% |
| | | | **stayed the same** | 0 | 0% | 0 | 0% | 0 | 0% | 0 | 0% | 0 | 0% | 0 | 0% |
| | | | **improved slightly** | 1 | 100% | 0 | 0% | 0 | 0% | 1 | 100% | 0 | 0% | 0 | 0% |
| | | | **improved considerably** | 0 | 0% | 0 | 0% | 0 | 0% | 0 | 0% | 0 | 0% | 0 | 0% |
| Total | 246 | 100.00% | | | | 48 | | 57 | | 34 | | 64 | | 34 | |

At the same time, there is evidence here that certain compliance mechanisms do make a difference. Among the cases featuring the issuance of notices of violation, 25.5 percent produced considerable improvement and another 31.9 percent led to slight improvement in the problem. Where considerable improvement occurred, moreover, the coders judged the regime to have either a very strong or a significant causal impact in two-thirds of the cases. The results are even more striking in the case of capacity building to enhance compliance. Among these cases, the problem improved slightly in 43.9 percent of the cases and considerably in 22 percent of the cases. The coders assessed the causal role of the regime as very strong or significant in 44.4 and 33.3 percent of the cases respectively. The data relating to the granting of transition periods suggest a slightly different pattern. When the problem improved considerably (which it did in 40.9 percent of the cases), the effect of the regime was very strong or significant in almost 90 percent of the cases. But in the absence of considerable improvement, the granting of a transitional period made little contribution to problem solving. These relationships do not add up to a clearcut picture regarding the links between the establishment of compliance mechanisms and the effectiveness of environmental regimes. But they do suggest that there are some specific relationships that make a difference in these terms.

The fact that arrangements involving capacity building constitute an important part of the picture regarding the links between compliance mechanisms and problem solving makes it all the more interesting to turn at this point to a consideration of broader questions regarding what have become known as the enforcement and the management approaches to compliance (Chayes and Chayes 1995). What can we learn from the IRD about the links between the selection of an overarching approach to compliance and both goal attainment and problem solving? Tables 4.7 and 4.8 present the data relating to this question.

The IRD data tell a dramatic story in this realm. In the query linking approaches to compliance with goal attainment, the coders concluded that the management approach is dominant in 94.1 percent of the cases. The parallel figure in table 4.8 linking approaches to compliance and problem solving is 89.4 percent of the cases. Those who create and administer international regimes exhibit an overwhelming preference for

**Table 4.7**
Compliance approach and goal attainment

| Compliance approach | Count | % | Goals fulfilled | | Degree of causality | | | | | |
|---|---|---|---|---|---|---|---|---|---|---|
| | | | | | Little or no causal impact | | Modest causal influence | | Large causal influence | |
| Enforcement approach | 24 | 5.94% | Yes | | **1** | 7.69% | **4** | 30.77% | **8** | 61.54% |
| | | | 13 | 54.17% | | | | | | |
| | | | No | | **4** | 36.36% | **1** | 9.09% | **6** | 54.55% |
| | | | 11 | 45.83% | | | | | | |
| Management approach | 380 | 94.06% | Yes | | **7** | 2.14% | **82** | 25.08% | **238** | 72.78% |
| | | | 327 | 86.05% | | | | | | |
| | | | No | | **17** | 32.08% | **22** | 41.51% | **14** | 26.42% |
| | | | 53 | 13.95% | | | | | | |
| Totals | 404 | 100.00% | 404 | | 29 | | 109 | | 266 | |

**Table 4.8**
Compliance approach and problem change

| Compliance approach | Count | % | Problem change | | | Problem change causal | | | | | | | | | |
|---|---|---|---|---|---|---|---|---|---|---|---|---|---|---|---|
| | | | | | | Little or no causal influence | | Modest causal influence | | Balanced causal influence | | Significant causal influence | | Very strong causal influence | |
| Enforcement approach | 15 | 10.57% | worsened considerably | 2 | 13.33% | 1 | 50% | 0 | 0% | 0 | 0% | 1 | 50% | 0 | 0% |
| | | | worsened slightly | 1 | 6.67% | 1 | 100% | 0 | 0% | 0 | 0% | 0 | 0% | 0 | 0% |
| | | | stayed the same | 0 | 0% | 0 | 0 | 0 | 0% | 0 | 0% | 0 | 0% | 0 | 0% |
| | | | improved slightly | 7 | 46.67% | 0 | 0 | 1 | 14.29% | 5 | 71.43% | 1 | 14.29% | 0 | 0% |
| | | | improved considerably | 5 | 33.33% | 0 | 0 | 0 | 0% | 2 | 40% | 3 | 60% | 0 | 0% |

| | | | | | | | | | | | | | | | |
|---|---|---|---|---|---|---|---|---|---|---|---|---|---|---|---|
| Management approach | 138 | 89.43% | worsened considerably | 10 | 7.97% | 2 | 27.27% | 6 | 54.54% | 0 | 0% | 1 | 9.09% | 1 | 9.09% |
| | | | worsened slightly | 21 | 15.22% | 11 | 52.38% | 6 | 28.57% | 2 | 9.52% | 2 | 9.52% | 0 | 0% |
| | | | stayed the same | 30 | 21.74% | 2 | 6.67% | 9 | 30.00% | 2 | 6.67 | 13 | 43.33% | 4 | 13.33% |
| | | | improved slightly | 37 | 26.81% | 4 | 10.81% | 7 | 18.92% | 6 | 16.22% | 11 | 29.73% | 9 | 24,32 |
| | | | improved considerably | 39 | 28.26% | 8 | 20.51% | 4 | 10.26 | 8 | 20.51 | 13 | 33.33 | 6 | 15.38 |
| Total | 153 | 100% | | | | 30 | | 33 | | 25 | | 45 | | 20 | |

procedures featuring management in contrast to enforcement. This is perhaps understandable if we bear in mind that international society is a society of states in which individual members are sensitive about any initiative that imposes restrictions on their sovereign rights. But it is not necessarily good news from the point of view of those striving to create international regimes that prove successful in fulfilling goals or solving problems.

What can we infer from the data regarding the links between approaches to compliance and both the attainment of goals and the amelioration of problems? The choice of a management approach is associated with the attainment of goals in 86.1 of the records in this query. Perhaps more impressive is the fact that the coders judged the regimes to have a large causal influence in 72.8 percent of the cases featuring the adoption of a management approach and the fulfillment of goals. Although the number of cases featuring an enforcement approach to compliance is small, the contrast in terms of results is significant. An enforcement approach leads to goal attainment in 54.2 percent of the cases, and among those cases the coders deemed the regime to have a large causal influence in 61.5 percent of the cases.

As we would expect, it is harder to solve problems than to attain stated goals. Still there is clear evidence in the database that regime design does matter. Of the 89.4 percent of the cases featuring a management approach, there has been considerable improvement or slight improvement in 26.9 and 21.8 percent of the cases respectively. Where considerable improvement has occurred, the coders judged the regime to have had a very strong causal impact in 15.4 percent of the cases and a significant causal impact in another 33.3 percent of the cases. Among the small number of cases in which an enforcement approach is dominant, considerable improvement occurred in 33.3 percent of the cases. But the coders did not treat the regime as having a strong causal influence in any of these cases. This evidence suggests that there are good reasons for selecting a management approach over an enforcement approach, at least when the problem at hand is environmental in character.

Overall, the data from the IRD suggest that regimes can make a difference, even when they lack the coercive capacity we often assume is needed to produce effective governance. This observation may be construed as compatible with a collective-action perspective on international regimes because efforts aimed at capacity building and initiatives featuring transitional periods are likely to influence the utilitarian calculations of a good many regime members. In our judgment, however, these findings lend even stronger support to social-practice perspectives on regimes. The emphasis on capacity building and the reliance on a management approach seem to reflect a strategy designed to draw individual regime members into an increasingly dense network of relationships that give rise to a situation in which compliance becomes an automatic response rather than a matter requiring an assessment of costs and benefits on a case-by-case basis.

## 4.4 Conclusion: Regime Effectiveness and Institutional Design

What inferences can we draw from this analysis about the determinants of institutional effectiveness and especially about procedures that will prove helpful to those responsible for creating or administering environmental regimes? In this concluding section, we return to the concerns about decision making and compliance articulated in the opening paragraph of this chapter. While they are not in any way definitive, the results we have reported should prove generally encouraging from the point of view of those who take an interest in the challenge of meeting the demand for governance in a stateless society.

With respect to decision rules, the principal message concerns the dominant role of consensus and—to a lesser extent—unanimity in the operation of international regimes. With regard to decision rules in practice, tables 4.2 and 4.4 show that consensus emerges as the primary decision rule in 58.4 percent of the records dealing with goal attainment and 53.8 percent of the records dealing with problem change. The only other decision rule that looms large in these tables is unanimity—21.3 percent of the records in table 4.2 and 20.5 percent of the records in table 4.4.

Some commentators may be inclined to treat these observations as confirming their fears regarding the limitations of international regimes as governance systems. But this inference is not warranted by the data included in the IRD. With respect to goal attainment, 77.9 percent of the cases featuring some form of consensus resulted in the fulfillment of goals. Even more important, the coders judged the regime to have had a large causal influence in 73.8 percent of the cases where goals were fulfilled. Nor is this finding about the results of employing consensus rules confined to the realm of goal attainment. Regimes relying on consensus rules produced improvement with regard to the problem—considerable or slight—in almost 50 percent of the cases. In cases featuring considerable improvement, the coders judged the regime to have had a significant or very strong causal influence in over 70 percent of the cases.

Explaining this collective preference for consensual procedures is not difficult. In a world of sovereign states, individual members of international regimes are seldom prepared to accept or abide by decisions made by others as a matter of course. But the interesting observation in the context of this analysis is that the search for consensus need not become an exercise in futility. Consensus, it should be emphasized, is an elastic concept. It certainly does not mean that all the members of a group must be ready and willing to endorse a measure with enthusiasm. On the contrary, consensus decision making is compatible with situations in which individual members abstain from voting, give in to pressure exerted by others, or acquiesce in the expectation that others will reciprocate with support for their issues at some future time. Building consensus, moreover, is a process that offers substantial scope for the exercise of entrepreneurial leadership. The record of environmental diplomacy is replete with cases in which skillful individuals (e.g., Tommy Koh in the case of the law of the sea or Mustafa Tolba in the cases of ozone and biodiversity) are able to devise formulas and broker deals that the members of a group eventually accept as an alternative to taking responsibility for the failure of consensual decision making. The real take-home message here is that the politics of processes featuring the creation of maximum-winning coalitions are fundamentally different than the politics of building minimum-winning coalitions. Whether or not one of these processes generates more effective results than the other is anything but clear. What is undeniable, however, is that there is little to be gained from an attempt to understand the use of consensual procedures in addressing international environmental problems by using familiar models dealing with coalition formation in domestic situations where participants generally strive to create minimum-winning coalitions (Riker 1962).

In the case of compliance mechanisms, our story is somewhat similar. The simple fact is that international environmental regimes do not rely heavily on coercive procedures in order to elicit compliance on the part of their members. The data included in tables 4.5 and 4.6 do show that regimes often make use of procedures featuring the issuance of notices of violation. But such notices seldom lead to the imposition of sanctions in the normal sense of the term. Beyond this, environmental regimes exhibit a pronounced tendency to favor the use of capacity building to enhance compliance and to focus on the development of transitional arrangements designed to encourage subjects—usually developing countries—to initiate processes that will lead to good records of compliance at a later stage.[6]

More generally, the data show that environmental regimes are far more likely to rely on a management approach to compliance than on an enforcement approach. The picture in this connection is slightly distorted by the fact that the IRD construes enforcement as a matter of using threats and punishments to attain results; mechanisms featuring promises and rewards are not coded as cases of enforcement. Nonetheless, the evidence in tables 4.7 and 4.8, which show that a management approach is adopted in 94.1 percent of the records dealing with goal attainment and 89.4 percent of the records dealing with problem change, is dramatic. Enforcement is simply not a prominent feature of efforts to promote compliance with international environmental regimes.

Is this cause for despair regarding the effectiveness of international regimes? As in the case of decision rules, the evidence included in the IRD does not support such a pessimistic conclusion. Reliance on capacity-building measures, for instance, leads to the fulfillment of goals in 89 percent of the cases, and the coders judged the causal influence of the regime to be large in 35.6 percent of these cases. The comparable figures for problem solving show that capacity building leads to improvement (considerable or slight) in almost two-thirds of the cases and that the causal role of the regime is very strong in at least a third of these cases. The findings are even more striking with regard to the reliance of regimes on a management approach to compliance. Regimes that rely on a management approach succeed in fulfilling goals 86.1 percent of the time; these regimes have a large causal influence in 72.8 percent of the cases.

Certainly, these results offer no basis for complacency regarding the effectiveness of environmental regimes. It is not easy to create effective regimes, and initial successes must not be allowed to lead to a decline in vigilance over time in such matters. Nevertheless, the evidence we have presented in this chapter suggests that there is no reason to dismiss international regimes because of their reliance on consensual decision-making procedures and their tendency to avoid familiar enforcement mechanisms. More generally, we are convinced that the logic of governance in world affairs—at least with regard to large-scale environmental issues—features processes that are not easy to understand on the basis of familiar models of domestic politics. But this is no reason to belittle their contribution to the supply of governance in international and transnational society.

# 5

# Programmatic Activities, Knowledge, and Environmental Problem Solving

## 5.1 Introduction

The knowledge available to manage environmental problems is always incomplete. Understanding of the dynamics of ecosystems is limited. Information about the biophysical mechanisms leading to environmental damage can be just as incomplete as our knowledge about human behavior affecting ecosystems. Similar gaps in knowledge plague efforts to create effective management regimes. Not surprisingly, therefore, actors often establish international institutions as part of a larger effort to enhance consensual knowledge about the nature of problems and the relative merits of alternative strategies for solving them.

Cognitivists argue that the growth of consensual knowledge can change the interests of states, making them more willing to accept the implementation of far-reaching international policies. They emphasize the role of learning as a response to the development of new scientific evidence as well as to the occurrence of external shocks.[1] Regimes, on this account, are frameworks encouraging states and other actors to pool their scientific resources in order to enhance understanding of cause-and-effect relationships and of the likely consequences of different policy options. Those who focus on factors of this sort take the view that regimes can become arenas that stimulate new thinking about transboundary issues and ways to solve them (Keohane 1988).

The key to this process lies in the role of programmatic activities carried out under the auspices of regimes. Programmatic activities involve organized efforts intended to support a regime's operation and to promote the implementation of its provisions. They can play a role in the creation of scientific knowledge, the emergence of insights regarding policies, the monitoring of implementation, the verification of compliance, or the handling of financial and technology transfers. These activities can help to reduce uncertainties and to build capacity in member states to implement the provisions of regimes effectively.

In this chapter, we focus on programmatic activities designed to produce consensual knowledge about the causes and effects of environmental problems and about the policies available to solve or manage them. For the most part, bodies designed specially for this purpose carry out such programmatic activities. The United Nations Framework Convention on Climate Change (UNFCCC), for instance, establishes two subsidiary bodies to handle these functions. The Subsidiary Body for Scientific and Technological Advice (SBSTA) deals with scientific and technical information relevant to the development of policies. The Subsidiary Body for Implementation (SBI) deals with matters of compliance and implementation review. The findings of the two bodies provide support for decision making on the part of the Conference of the Parties (Oberthür and Ott 1999).

What can the IRD tell us about the links between programmatic activities and the scope of consensual knowledge? Does success in problem solving depend on the growth of consensual knowledge regarding a specific environmental issue? To answer these questions, we proceed in this chapter as follows. Section 2 explores the nature of programmatic activities in greater detail and groups them into four broad categories. The next section focuses on theoretical considerations and develops a set of expectations about the links among programmatic activities, the production of knowledge, and problem solving. Sections 4 and 5 turn to the empirical evidence. Specifically, section 4 delves into the evidence regarding the links between programmatic activities and the production of consensual knowledge. Section 5 then goes on to probe the relationships between the growth of knowledge and success in efforts to solve problems.

## 5.2 Types of Programmatic Activities

A desire to obtain better information and more complete knowledge is one incentive for states to join international environmental regimes. In many cases, the aftermath of regime formation as such is dominated by efforts to get programmatic activities up and running. Strong support for such efforts arises from the fact that member states typically have a limited understanding of the causes and effects of problems, especially when these causes and effects are transboundary in nature. A typical problem arises from the limited capacity of developing countries, a situation that often leads more developed countries to provide support for the establishment of scientific and technical infrastructure in return for active engagement on the part of developing countries. In many cases, it makes sense as well to agree on a division of labor regarding monitoring and scientific research. Often, industrialized countries are able to take the lead in implementing specific regime functions because they already possess the scientific and technical capacity required.

A common rationale for creating regimes involves the intensification of scientific and technical collaboration in the relevant issue areas. A major goal embedded in the 1959 Antarctic Treaty, for example, is "to promote international cooperation in scientific investigation in Antarctica" (Article III). The adoption in 1979 of the Convention on Long-Range Transboundary Air Pollution (CLRTAP) led to a sizable expansion of scientific research dealing with acid precipitation. The Bucharest Declaration of 1985, which became the first important international agreement aimed at protecting the Danube River Basin from pollution, specifies that the participating countries will establish an international monitoring system for water quality at border-crossing points. The purpose of the Great Lakes Water Quality Agreement, first signed in 1972 and revised and upgraded in 1978, is "to restore and maintain the chemical, physical, and biological integrity of the waters of the Great Lakes Basin Ecosystem." To achieve this purpose, the United States and Canada agreed "to make a maximum effort to develop programs, practices and technology necessary for a better understanding of the Great Lakes Basin Ecosystem and to eliminate or reduce to the maximum extent practicable the discharge of pollutants into the Great Lakes System."[2]

Governments understand that there are mutual benefits associated with the development of common methodologies for assessing the causes and effects of environmental problems and the exchange of data pertaining to these matters. They expect similar gains from the development of common research plans, the creation of international programs that facilitate the monitoring of the environment or review the adequacy of commitments, and the exchange of information about the implementation of provisions of international regimes at the national level. Programmatic activities associated with specific regimes often benefit from scientific and technical knowledge arising from the efforts of intergovernmental organizations or international scientific associations. The World Meteorological Organization (WMO), for example, provides a framework for the conduct of research on weather and climate. In the process, the WMO has spawned global programs like the World Weather Watch Programme, the World Climate Research Programme, the Atmospheric Research and Environment Programme, and the Hydrology and Water Resources Programme. Environmental regimes can and often do take advantage of the knowledge produced by these programs (Soroos and Nikitina 1995).

Transnational networks among environmental scientists and experts on economic and legal matters contribute to the creation of knowledge useful in pursuing the goals of specific regimes. The World Conservation Union (IUCN), for instance, combines members from more than 100 government agencies and more than 800 NGOs. The organization provides policy advice and technical support to international institutions, plays a role in assessing sites nominated for world heritage status, participates in monitoring the condition of the world's species, and supports developing countries seeking to prepare new environmental legislation—to mention only a few of the contributions this organization makes to the work of international regimes. Other scientific bodies, such as the International Council for Science (ICSU) and the International Council for the Exploration of the Sea (ICES), make similar contributions.

The production of consensual knowledge in regimes is closely linked as well to scientific programs carried out on the national level. Although international programs structure scientific discourse on environmental problems and provide guidelines for the work of national or transnational scientific actors, the scientific and technical activities of national or transnational bodies are not oriented exclusively toward research agendas and functions agreed on at the international level. Their work frequently addresses issues that lie beyond the clearly specified frameworks of these international programs. Nonetheless, the findings of national programs often flow into transnational scientific discourses. States play an important role in this process. They endow international organizations with the financial resources required for implementing international research and monitoring programs. They contribute to the implementation of these programs by way of participation on the part of governmental or semigovernmental research institutes. They collaborate with private actors able to contribute scientific and technical expertise on the national and international levels.

The IRD distinguishes and asks coders for input regarding four types of programmatic activities relevant to the production of consensual knowledge and efforts to solve environmental problems.

### 5.2.1 Scientific Assessment and Monitoring of the Causes and Effects of Problems

Some regimes aim first and foremost to strengthen consensual knowledge usable as a basis for the development of international policies in subsequent stages. In this connection, assessment can produce findings that motivate states to expand or strengthen the provisions of existing regimes. International agreements dealing with environmental issues often mandate the establishment of institutionalized mechanisms dedicated to assessment and monitoring. The 1979 convention on acid precipitation, for instance, stipulates that regime members "shall by means of exchange of information, consultation, research and monitoring, develop . . . policies and strategies which shall serve as a means for combating the discharge of air pollutants" (Article 3). This convention also confirms and reinforces the importance of existing monitoring efforts conducted through the work of the Cooperative Programme for Monitoring and Evaluation of the Long-Range Transmission of Air Pollutants in Europe (EMEP) (Levy 1993). The 1992 Convention on the Protection of the Marine Environment of the Baltic Sea Area, which replaced an earlier 1972 treaty, stipulates that in order "to facilitate monitoring activities in the Baltic Sea Area the Contracting Parties undertake to harmonize their policies with respect to permission procedures for conducting such activities" (Article 24.1). In the same convention, states agree to "co-operate in developing inter-comparable observation methods, in performing baseline studies and in establishing complementary or joint programmes for monitoring" (Article 24.3). Although weak in other respects, the 1985 Vienna Convention for the Protection of the Ozone Layer calls on the parties to create an international legal framework to promote cooperation regarding systematic observation of the ozone layer and the monitoring of CFC production (Article 2). States and international organizations had taken some initial steps to use satellite, airborne, and ground-based systems to monitor stratospheric ozone as well as the chemical compounds and atmospheric conditions affecting its composition before the signing of the Vienna Convention. But the creation of the ozone regime has substantially strengthened scientific assessment and monitoring relevant to this problem (Roan 1989; Parson 2003).

### 5.2.2 Research Designed to Probe the Causes and Effects of the Problem

Research is closely linked to the assessment and monitoring activities carried out under the auspices of regimes.[3] Assessment, monitoring, and research are addressed in some agreements in a single article. In other cases, they are treated one after another in the texts of environmental agreements. Article 3 of the 1985 Vienna Convention, for example, builds on the general provisions relating to monitoring and research included in Article 2. Addressing the issue of systematic observation and research, it specifies that states shall "undertake, as appropriate, to initiate and co-operate in, directly or through competent international bodies, the conduct of research and scientific assessments." A list of major issues to be addressed by regime bodies (e.g., the physical and chemical processes that may affect the ozone layer, the human health and other biological effects stemming from modifications of the ozone layer, effects on the climate system, impacts of ozone depletion on natural and synthetic materials useful to humankind) follows this general provision. Programmatic activities—often carried out in conjunction with national research efforts—play key roles in the collection of data and the development and implementation of large-scale research plans, whether they are financed by governments, intergovernmental organizations, or private funders. The Antarctic Treaty System illustrates how research networks can grow over time. The Scientific Committee on Antarctic Research (SCAR), operating as an ICSU body, provides independent scientific advice to the Antarctic Treaty System and is expected to initiate, promote, and coordinate scientific research in Antarctica. SCAR relies heavily in turn on national research programs, such as the U.S. Antarctic Research Program (Joyner 1998). Similarly, national weather services, climate research efforts, and space programs all coordinate research relating to atmospheric and climatic matters. Their findings become major inputs in the scientific discourses that arise in regimes for the protection of the ozone layer and the global climate system.

### 5.2.3 Systems of Implementation Review

As the ranks of environmental regimes have grown, implementation review has become a prominent issue. In many regimes, the review of implementation was handled at first on an ad hoc basis or even ignored completely. The Baltic Sea regime illustrates this situation. Initially, the East-West conflict constrained efforts to establish regular procedures for the submission of information about implementation carried out at the domestic level. But the end of the Cold War and replacement of the 1974 Helsinki Convention by a renegotiated convention for environmental protection in the area of the Baltic Sea in 1992 brought about a situation in which "reporting and implementation review activities have...expanded and intensified" (Greene 1998, 191). By 1985, the Vienna Convention could stipulate that the "Conference of the Parties shall keep under continuous review the implementation of this Convention" (Article 6). Systems of implementation review (SIRs) can certify that measures deemed necessary to fulfill the obligations of a regime are actually carried out (Victor et al. 1998). In the process, they can limit the occurrence of deliberate cheating and build trust among regime members (Marauhn 1996; Underdal 1998). The availability of information about the success or failure of efforts on the part of member states to implement regime requirements within their own jurisdictions can make it possible to respond to emerging compliance problems at an early enough stage to prevent conflicts over such matters from undermining a regime. But SIRs can do more than reduce uncertainty about the behavior of other regime members. Handled properly, they can inform regime members about the effectiveness of existing measures and indicate whether additional or alternative measures are needed to address the problem at hand. SIRs typically rely on reports submitted by individual regime members. Under the circumstances, it is important to recognize that the submission of national reports is often delayed and that individual members can encounter serious problems in collecting the information required or prove unwilling to submit data to regime bodies. It is hardly surprising, therefore, that regimes often devote time and energy to solving problems relating to the submission of national reports and to building capacity in member states for the collection of data on implementation.

### 5.2.4 Review of the Adequacy of Existing Commitments

Experience with implementation review can raise questions about the adequacy of the commitments included in a regime's constitutive provisions. Review processes can also examine whether there is a need to redesign a regime's provisions to bring them into line with recent technological, economic, legal, or political developments. Some regimes have provisions covering the review of existing commitments built into the conventions or treaties under which they operate. But as the case of the Antarctic Treaty makes clear, these provisions raise sensitive issues and are often hard to activate. In other cases, opportunities for the review of existing commitments arise as experience with a regime grows. It has become apparent, for instance, that the provisions of the Kyoto Protocol on climate change will need to be revisited in conjunction with the end of the first commitment period. The ozone regime provides a particularly interesting example in this connection. It establishes different procedures for accelerating phaseout schedules for chemicals already covered, in contrast to making decisions regarding proposals to add chemicals to the regulated list. The Conference of the Parties operating under the terms of the 1987 Montreal Protocol was able to accelerate initial phaseout schedules largely on the basis of assessment reports and without triggering a need for ratification on the part of member states. Decisions about the addition of new chemicals (e.g., methyl chloroform) to the list of those scheduled for phaseout, on the other hand, require a more thorough review process together with explicit consent on the part of member states. In both cases, however, assessments of the adequacy of existing commitments in the light of new scientific evidence and experience with the implementation of the provisions of the regime have played a key role in the remarkable evolution of this regime during the period since its inception (Litfin 1994; Parson 2003).

## 5.3 Theory—Programmatic Activities, Knowledge, and Problem Solving

Power- and interest-based approaches dominated the early stages of research on international regimes (Krasner 1983a; Hasenclever, Meyer, and Rittberger 1997). These approaches direct attention to the international level. They explain the rise and fall of individual regimes in terms of considerations of structural power and various perspectives on problem structure. Over time, the limitations of these approaches have become increasingly apparent. They offer static views of institutions. They cannot explain situations in which institutions remain effective despite changes in the distribution of power among participating states. They do not account for changes in national preferences that transform the character of collective-action problems. Mainstream rational-choice analyses stress the importance of assuming that the preferences of states are fixed as a precondition for constructing models of collective action. More recently, however, interest-based approaches have had to consider the roles that nonstate actors play in the formation of state preferences and the ways in which the emergence of new norms and the evolution of knowledge can alter the interests of states (Snidal 2002).[4]

In some respects, a knowledge-based approach to the study of regime effectiveness complements the interest-based approach. The complexity of transboundary environmental issues makes it essential to come to grips with uncertainties in efforts to establish clearcut preferences among policy options. Knowledge-based analyses start from the assumptions that institutions can operate in ways that change both the content of existing knowledge and the extent to which it is consensual in nature and that these changes often affect the preferences of individual states. Critical to this way of thinking about environmental regimes is the premise that changes in prevailing knowledge can lead states to accept the implementation of more far-reaching provisions designed to solve specific environmental problems.

The connections between knowledge and environmental problem-solving are highly complex. Many forces are involved in the production of knowledge; we cannot assume that regimes will play a dominant role in this realm. Governments with great scientific, technological, and financial resources can influence the production of consensual knowledge regardless of what others do and make use of the resultant knowledge to influence the development of discourses associated with specific regimes. Against this background, we have sought to explore the roles that international regimes and, more specifically, the implementation of programmatic activities play in the creation of consensual knowledge (Ebbin 2004). The premise of knowledge-based approaches is that the more regimes engage in assessment, monitoring, scientific research, implementation review, and reviews of the adequacy of existing commitments, the more likely improvements in the quality of consensual knowledge become. Conversely, a failure to engage in such activities will be reflected in a lack of progress toward the development of consensual knowledge. Testing this idea requires an effort to demonstrate not only a clear connection between programmatic activities and the growth of consensual knowledge but also a link between the absence of such activities and a failure of consensual knowledge to grow.

This line of analysis treats programmatic activities as *independent variables* that influence the growth of consensual knowledge, treated as the *dependent variable*. The theoretical stance embedded in this argument is not limited to a conception of regimes as intervening variables that affect the connection between power or interests on the one hand and outcomes or impacts on the other. A consideration of regime attributes as independent variables implies that various outcomes and impacts could not occur in international society in the absence of effective regimes. This view of regimes accords these arrangements a more fundamental role than the role envisioned in the early theoretical work on regimes (Krasner 1983b).

So far, our discussion about the contribution of regimes to the production of consensual knowledge has focused on impacts that occur on the level of the regime as a whole. But a cognitivist approach to the study of regimes also leads us to expect that the impacts of programmatic activities will occur at the level of member states. Information about both the causes and effects of environmental problems and policies for solving them may be unevenly distributed among important actors. States that are poorly informed about such matters will be more dependent than others on a regime's programmatic activities to remedy this knowledge deficit. Assuming that important states are generally better informed than others, an exclusive focus on important states in the analysis of regime consequences runs the risk of undervaluing the roles that programmatic activities play in the production of relevant knowledge. Of course, other factors sometimes account for changes in views regarding cause-and-effect relationships or in thinking about policy options. States can mobilize their own scientific and technical resources to advance their understanding of these matters. Domestic as well as transnational nonstate actors can provide support for the creation of consensual knowledge and its dissemination to decision makers or the general public.

Some analysts have paid special attention to the role of epistemic communities treated as networks that include scientific, technical, legal, and policy experts who share common views about cause-and-effect relationships, common conceptions of the goals to be fulfilled by policymakers, and common preferences regarding the choice of specific policies for problem solving (Adler and Haas 1992). But other cognitivists pay less attention to the existence of such communities. They emphasize the role of other entities, including nonprofit organizations, scientific networks, international nongovernmental organizations, and economic actors that are open-minded about the implementation of environmental policies, whose activities can stimulate social learning on the part of important states (Keck and Sikkink 1998; Social Learning Group 2001a, 2001b). Institutions cannot achieve desired outcomes without the support of many actors. On the other hand, these actors do not always pull in the same direction; they can promote or impede the evolution of consensus regarding cause-and-effect relationships. While the work of scientists associated with regimes normally features rational discourse, the relationship between science and policy can and often does become politicized in efforts to solve specific environmental problems. Several members of the European Communities, for example, raised serious doubts during the first half of the 1980s concerning the validity of the theory underlying the views of atmospheric chemists regarding the impacts of chlorine on the stratospheric ozone layer (Benedick 1991). It took a large-scale international effort in the realm of scientific monitoring and research to persuade European states to give up their resistance to the adoption and implementation of policies designed to phase out these substances. A similar debate took place in the United States in the 1990s, when many policymakers opposed the acceptance of the Kyoto Protocol on the grounds that scientific knowledge was not sufficiently developed to justify such a response to the prospect of climate change.

A finding that this line of analysis regarding the role of programmatic activities in the production of knowledge is supported empirically would constitute a basis for cautious optimism regarding the effectiveness of international environmental regimes. It would also help to set the stage for an assessment of the roles regimes play in solving problems. New consensual knowledge can structure the way policymakers think about the consequences or the appropriateness of different approaches to problem solving. Such processes are determined more by arguing than by the exercise of power (Risse 2000). As a result, they enhance the influence of scientists as well as technical, legal, or policy experts and produce choices among alternative approaches to problem solving that differ from those predicted by power-based or interest-based models. Yet these processes are not without problems of their own. The growing influence of experts may give rise to problems of accountability and, in the final analysis, generate a significant democracy deficit. But there is no denying that programmatic activities may entrain processes whose consequences are felt far beyond the production of consensual knowledge as such.

Before we turn to the IRD to generate empirical data regarding these matters, it may help to summarize the theoretical expectations arising from this account of programmatic activities, knowledge, and problem solving. The role of knowledge is not limited to processes of regime formation. Consensual knowledge can change the way actors think about problems, broaden the scope of the norms they adopt, and influence the choice of strategies designed to solve environmental problems. It is therefore critical to understand the roles that regimes play in the development of consensual knowledge. Based on cognitivist perspectives, we have explored the links between the programmatic activities regimes undertake and the growth of knowledge about both the nature of the relevant problems and the relative merits of different approaches to problem solving. Thus, we turn to the IRD in search of evidence for a two-stage model in which programmatic activities influence the content and status of consensual knowledge, and knowledge in turn affects the selection of strategies and their effectiveness in solving specific problems.

## 5.4 Empirical Evidence—Programmatic Activities and Knowledge

To test the theoretical expectations sketched in the preceding section, we can develop queries that make use of the data included in the IRD. Information about the existence and nature of programmatic activities is included in the section of the IRD Data Protocol on regime attributes. Although the protocol lists nine separate programmatic activities, it devotes particular attention to the four described in the preceding sections, because they figure prominently in analytic debates about the value of knowledge-based approaches to regime analysis.

A few conceptual issues require clarification before we turn to the data. How can we measure the evolution of consensual knowledge? At what level should we tackle this issue? In this chapter, we employ two measures to characterize the state of consensual knowledge. First, we ask whether *the nature of the problem* is well understood. Consensus in this realm is a function of the level of agreement about the nature, causes, and consequences of the problem. Second, we ask about the completeness of *information about the options available for dealing with the problem*. Strategies for tackling a problem may reflect different ways of framing the issue at hand or focus on measures taken within different sectors. For purposes of analysis, we approach these variables on two levels: the level of the regime element, and the level of individual actors and, especially, important states.

On both levels, measurements of consensual knowledge are affected by the conclusions articulated in the precoding agreement. As we observed in chapter 2, we treat regimes as problem-driven governance systems (see also table 2.1). The data characterizing levels of understanding about the nature of the problem and about policy options for solving the problem are evaluated with regard to the problem as framed in the precoding agreement. Most of the regime elements address a single problem. For some regime elements, however, we ended up focusing on two or three problems coded separately by our expert coders. As a result, the data describing the extent of knowledge about the nature of the problem as well as information about policy options normally exceed the total number of regime elements.

Taking these considerations into account, we developed the following queries concerning the links between programmatic activities and the status of consensual knowledge:

*Query 1* Are there clearcut links between a regime's programmatic activities and the level of understanding regarding the nature of the problem?

*Query 2* Are there clearcut links between a regime's programmatic activities and the level of information about the options available for dealing with the problem?

*Query 3* Did understanding about the nature of the problem change on the level of all states in the issue area and how much of this change is attributable to the operation of the regime?

*Query 4* Did understanding about the nature of the problem change within important member states and how much of this change is attributable to the operation of the regime?

*Query 5* Did information about the options available for tackling the problem change on the level of all states in the issue area and how much of this change is attributable to the operation of the regime?

*Query 6* Did information about the options available for tackling the problem change within important member states and how much of this change is attributable to the operation of the regime?

### 5.4.1 Effects on the Level of Regime Elements

The database measures levels of understanding on a five-point scale. The upper and lower boundaries are defined as understanding *very strongly established* or *not at all established*. The top category of understanding covers situations where there is overall consensus regarding the nature, causes, and consequences of a problem as well as goals and solutions in the issue area. The bottom category on this ordinal scale covers the opposite conditions. The middle of the scale covers situations involving partial consensus, including agreement on some but not all factors or growing but not yet full consensus. Table 5.1 summarizes the data regarding links between programmatic activities and the level of understanding about the nature of the problem.

Strongly established understanding emerged in about two-thirds (68.3 percent) of the 205 records included in this table. These records contain data pertaining to 150 of the 172 regime elements included in the IRD. In some ways, this finding is encouraging. At least strongly established understandings regarding the nature of the problem exist with regard to the majority of transboundary environmental problems. Yet very strongly established understanding exists in only 28.3 percent of the cases, and insufficient understanding occurs in almost a third (31.7 percent) of the problems included in the table.

What about the links between level of understanding of the nature of the problem and the operation of programmatic activities? Programmatic activities focusing on assessment and monitoring are in place for more than half of these cases; activities featuring research on cause-and-effect relationships are also in place in more than half of the cases. Reviews of implementation and of the adequacy of commitments occur in nearly half of the cases. Monitoring activities are underway in a large majority of the cases featuring low established understanding, and programs for research on cause-and-effect relationships are present in nearly two-thirds of these cases. More than half of the 140 cases featuring strongly or very strongly established understanding include programmatic monitoring. Nearly half of these cases feature research programs dealing with cause-and-effect relationships. Overall, the data in table 5.1 show that at least one of the programmatic activities covered in this chapter is in place in all the cases included. Of course, this does not tell us if there are clear links between the operation of programmatic activities and the level of understanding of the problem established. Rather, it suggests the following puzzle. Why is there no straightforward association between variance in the dependent variable—level of understanding of the nature of the problem—and the number and character of the programmatic activities in place?

**Table 5.1**
Programmatic activities and understanding about the nature of the problem

| | | | PROGRAM_ACTIVITIES (RA 16) | | | | | | | | |
|---|---|---|---|---|---|---|---|---|---|---|---|
| | | Total of problems for which correlation can be made between forms RF22/RA16 | 1 | | 2 | | 5 | | 8 | | 10 = At least one of the four scientific programs (1, 2, 5 or 8) existed |
| PROBLEM_UNDERSTAND (RF22) | 1 | 58 (28.3%) | 23 | 17.3% | 15 | 13.4% | 13 | 13.4% | 23 | 24.2% | 40 (69.0%) |
| | 2 | 82 (40.0%) | 61 | 45.9% | 54 | 48.2% | 49 | 50.5% | 40 | 42.1% | 67 (81.8%) |
| | 3 | 41 (20.0%) | 33 | 24.8% | 34 | 30.4% | 28 | 28.9% | 25 | 26.3% | 37 (90.2%) |
| | 4 | 24 (11.7%) | 16 | 12.0% | 9 | 8.0% | 7 | 7.2% | 7 | 7.4% | 20 (83.3%) |
| | 5 | 0 (0.0%) | 0 | 0% | 0 | 0% | 0 | 0% | 0 | 0% | 0 (0%) |
| **Total** | | **205 (100.0%)** | **133** | **100%** | **112** | **100%** | **97** | **100%** | **95** | **100%** | **164 (80%)** |

PROGRAM_ACTIVITIES (Programmatic Activities Established in Regime Elements): 1 = scientific monitoring of causes and effects of the problem, 2 = research about causes and effects of the problem, 5 = review of implementation, 8 = reviewing adequacy of commitments, 10 = at least one of the four scientific programs existed
PROBLEM_UNDERSTAND (Level of Understanding about the Nature of the Problem): 1 = very strongly established understanding, 2 = strongly established understanding, 3 = partially established understanding, 4 = low established understanding, 5 = understanding not at all established

Before tackling this question, we turn to the data relating to our second query dealing with the links between programmatic activities and the completeness of information about available policy options. Here the scale runs from *very high completeness* in cases where all relevant information about policy options is available through *medium completeness* in cases where information about at least some of the policy options is inadequate to *low completeness* in cases where important information about options is unavailable and some potential options may not even be identified. Table 5.2 summarizes the data regarding this query.

As the table indicates, medium completeness of knowledge about policy options is present in over 60 percent of the cases, low completeness occurs in 20 percent of the cases, and very high completeness occurs in 19.4 percent of the cases. These findings are broadly similar to those regarding the links between programmatic activities and understanding of the nature of the problem. A range of programmatic activities aim to provide information about the nature and consequences of policy options. But differences in the dependent variable—completeness of information about policy options—are not closely linked to the operation of programmatic activities.

These queries suggest the need for caution in thinking about the links between programmatic activities and knowledge. There is some indication that a dense network of programmatic activities can influence understanding of the nature of the problem as well as the nature and availability of policy options. But in many cases, the extent of programmatic activities is roughly the same in cases where understanding and completeness are high and cases where these measures of knowledge are low. In this context, it is important to consider the role of intervening variables such as the complexity of the issues that regimes address. Some problems are more complex than others, both in biophysical terms and in terms of the character of the relevant social context. Ideally, we should strive to control for variance in these terms in evaluating the links between programmatic activities and the status of knowledge about both the nature of the problem and the policy options available to address it.

The relationships emerging from tables 5.1 and 5.2 pertain to the impacts programmatic activities have during the evolution of regime elements. Improving information about cause-and-effect relationships and about policy options takes time. In many regimes, the occurrence of

**Table 5.2**
Programmatic activities and the completeness of information about policy options

| | | | PROGRAM_ACTIVITIES (RA 16) | | | | | | | | | |
|---|---|---|---|---|---|---|---|---|---|---|---|---|
| INFORMATION_OPTIONS_PROBLEM (RF23) | | Total of problems Coded | 1 | | 2 | | 3 | | 4 | | 5 | |
| | 1 | 35 (19.4%) | 26 | 20.3% | 20 | 18.7% | 24 | 16.2% | 18 | 18.4% | 18 | 19.8% |
| | 2 | 109 (60.65) | 77 | 60.2% | 67 | 62.6% | 99 | 66.9% | 62 | 63.3% | 55 | 60.4% |
| | 3 | 36 (20.0%) | 25 | 19.5% | 20 | 18.7% | 25 | 16.9% | 18 | 18.4% | 18 | 19.8% |
| **Total** | | 180 (100%) | **128** | 100% | **107** | 100% | **148** | 100% | **98** | 100.1% | **91** | 100% |
| | | | | | | | | | | | | |
| INFORMATION_OPTIONS_PROBLEM (RF23) | | | 6 | | 7 | | 8 | | 9 | | 10 = At least one of the nine programs existed | |
| | 1 | | 22 | 22.2% | 7 | 14% | 15 | 16.3% | 14 | 14.1% | 34 | 97.1% |
| | 2 | | 56 | 58.9% | 33 | 66% | 62 | 67.4% | 69 | 69.7% | 108 | 99.1% |
| | 3 | | 17 | 17.9% | 10 | 20% | 15 | 16.3% | 16 | 16.2% | 34 | 94.4% |
| **Total** | | 180 (100%) | **95** | 100% | **50** | 100% | **92** | 100% | **99** | 100% | **n.a.** | **n.a.** |

PROGRAM_ACTIVITIES (Programmatic Activities Established in Regime Elements): 1 = scientific monitoring of causes and effects of the problem, 2 = research about causes and effects of the problem, 3 = expert advice, 4 = compliance monitoring, 5 = review of implementation, 6 = verification of compliance, 7 = financial and technology transfer, 8 = reviewing adequacy of commitments, 9 = information management, 10 = at least one of the four scientific programs existed

INFORMATION_OPTIONS_PROBLEM (Completeness of Information about Policy Options): 1 = very high completeness, 2 = medium completeness, 3 = low completeness temporal watersheds allows us to compare developments occurring during earlier and later stages in the lives of regimes. Thus, we can ask whether consensual knowledge improves in the aftermath of a temporal watershed, and we can observe such developments in a number of regimes. Understanding of problems like "the [vulnerable] state of the Baltic Sea," or "prevention/control of pollution from (emissions of) hazardous substances and nutrients by up- and downstream countries into the aquatic environment of the Danube River Basin and into the Black Sea," or "protection of endangered species" reached higher levels during later stages in the development of regimes for the protection of the Baltic Sea and the Danube River Basin and for the management of trade in endangered species.

We may also ask about the growth of knowledge in the absence of programmatic activities: Are there cases exhibiting high levels of consensual knowledge in the absence of programmatic activities? There are a few cases in which strongly or partially established understanding regarding the nature of the problems prevails without the benefit of assessment, monitoring, or research. For the most part, however, these are cases in which such tasks were carried out by other bodies, including intergovernmental organizations and national research programs. Understanding regarding the nature of the problem of "coastal oil pollution and sea-bird deaths due to oil pollution from intentional discharges and accidental oil pollution"—the central concern of regulations promulgated under OILPOL and MARPOL—has been strongly established since the 1950s. Accordingly, the regime has focused greater attention on monitoring the behavior of shipping companies and flag states with regard to the implementation of relevant regulations than on improving understanding of the nature of the problem. In effect, the fact that the problem was already well understood has allowed this regime to concentrate on matters such as verifying compliance, reviewing implementation, and facilitating financial and technology transfers (Mitchell 1994).

### 5.4.2 Effects on the Level of Important States

The IRD data suggest that programmatic activities do play a role in the growth of knowledge. But so far, we have been unable to demonstrate clearcut links between the operation of specific programmatic activities and levels of understanding and completeness of knowledge. In this regard, we anticipate that the next set of queries will produce findings that are more informative. Queries 3–6 focus on the growth of knowledge at the level of member states and seek to tease out the causal role of programmatic activities in this realm. Two different measures are helpful in addressing this set of issues: measurements pertaining to knowledge of the nature of the problem and information regarding policy options on the part of all states, and parallel measurements focusing exclusively on important states.

Before we present the empirical findings regarding this topic, a few observations about measurement procedures are in order. We divided changes in the knowledge available to states into two categories: *little or no change* and *significant change*. In addition, our expert coders provided judgments regarding the roles regimes have played as determinants of changes in knowledge. In this connection, they worked with a three-point scale. *Little or no causal impact* describes cases in which nonregime factors explain most of the changes in knowledge. *Modest causal influence* occurs in cases where nonregime factors are still important, but the regime makes a difference with regard to changes in knowledge of the nature of the problem or the character of policy options. *Significant causal influence* means that the operation of a regime is more important than nonregime factors in determining changes in knowledge available to regime members.

Tables 5.3a and 5.3b present empirical findings based on data included in the IRD about these matters for all states. Significant changes in knowledge about both the nature of the problem and the availability of policy options occurred in just over half the cases included in the tables. When significant changes in knowledge occur, the regime is judged to have a significant causal influence over half the time. Modest causal influence occurs in roughly a third of the cases. The impact of the regime was insignificant in less than 10 percent of these cases. Of course, regimes cannot account for changes in knowledge with regard to all regimes or all time periods within individual regimes. Nonetheless, these findings regarding the roles regimes play in causing changes in the way problems are understood and in the information about policy options are striking.

The findings imply that programmatic activities do play a role in shaping the knowledge available to the members of regimes. Even so, we

**Table 5.3a**

Change in understanding regarding the nature of the problem on the level of all states

| | | PROBLEM_UNDERSTAND_CHANGE_CAUSAL (RC12) | | | | | | | |
|---|---|---|---|---|---|---|---|---|---|
| PROBLEM_ UNDERSTAND_ CHANGE (RC12) | Change in understanding regarding the nature of the problem in all states (Total) | 0 | | 1 | | 2 | | 3 | |
| 1 | 98 (48.3%) | 38 | 100% | 21 | 72.4% | 8 | 18.2% | 31 | 33.7% |
| 2 | 105 (51.7%) | 0 | 0% | 8 | 27.6% | 36 | 81.8% | 61 | 66.3% |
| **Total** | **203 (100.0%)** | **38** | **100%** | **29** | **100%** | **44** | **100%** | **92** | **100%** |

PROBLEM_UNDERSTAND_CHANGE_CAUSAL: 0 = not applicable (no change occurred), 1 = little or no causal impact, 2 = modest causal influence, 3 = significant causal influence

PROBLEM_UNDERSTAND_CHANGE (Change in Understanding regarding the Nature of the Problem in all States): 1 = little or no change, 2 = significant change

**Table 5.3b**
Changes in the completeness of information about policy options on the level of all states

| | | PROBLEM_UNDERSTAND_CHANGE_CAUSAL (RC12) | | | | | | | |
|---|---|---|---|---|---|---|---|---|---|
| PROBLEM_ UNDERSTAND_ CHANGE (RC12) | Change in understanding regarding the nature of the problem in all states (Total) | 0 | | 1 | | 2 | | 3 | |
| 1 | 98 (48.3%) | 38 | 100% | 21 | 72.4% | 8 | 18.2% | 31 | 33.7% |
| 2 | 105 (51.7%) | 0 | 0% | 8 | 27.6% | 36 | 81.8% | 61 | 66.3% |
| **Total** | 203 (100%) | 38 | 100% | 29 | 100% | 44 | 100% | 92 | 100% |

OPTIONS_PROBLEM_CHANGE_CAUSAL: 0 = not applicable (no change occurred), 1 = little or no causal impact, 2 = modest causal influence, 3 = significant causal influence
OPTIONS_PROBLEM_UNDERSTAND_CHANGE (Change in Information about Policy Options for a Problem in all States): 1 = little or no change, 2 = significant change

should be cautious in interpreting these results. Even in cases where regimes influence knowledge significantly, other actors are active. The occurrence of cases in which substantial changes in knowledge occur but regimes exert only a modest causal influence demonstrates that regimes are embedded in transnational knowledge networks. It is important to remember as well our previous finding that knowledge is very complete in only about 20 percent of the cases (table 5.2), even though significant changes in information about policy options occur in more than half the cases (table 5.3b). These observations suggest that it would be helpful to devote more energy to the production and dissemination of information about policy options.

Moving on now from a consideration of the set of all states to a focus on the subset of important states, our data point to similar conclusions. In most cases, the coders identified four to six states as important actors in the relevant regime. Tables 5.4a and 5.4b present the data regarding changes in knowledge and the role of regimes in bringing about such changes for the subset of important states. Significant changes occurred in slightly less than half the cases with regard both to knowledge about the nature of the problem and to information about policy options. But in cases where knowledge did change significantly, the regimes had a significant causal impact well over half the time; their impact was seldom negligible. Broadly speaking, then, the results are similar whether we consider all states or only the subset of important states.

These findings are encouraging. Still, the fact that the knowledge available to important states shows little or no change in over half the cases requires further consideration. What accounts for the lack of change in these cases? It may be that knowledge remains on a low level in cases where no significant changes are recorded. The IRD can neither confirm nor refute this pessimistic perspective. Of course, an alternative interpretation suggests that knowledge does not change in these cases because the relevant actors are well informed at the outset. But a desire to avoid whitewashing or rushing to overly optimistic judgments leads us to remain skeptical about this interpretation.

A look at developments in specific important states can amplify the analysis presented in the preceding paragraphs (tables 5.4c and 5.4d). It is clear, to begin with, that significant changes in knowledge were not

**Table 5.4a**
Change in understanding about the nature of the problem on the level of important states

| | | PROBLEM_UNDERSTAND_CHANGE_CAUSAL (RC12) | | | | | | | |
|---|---|---|---|---|---|---|---|---|---|
| PROBLEM_ UNDERSTAND_ CHANGE (RC12) | Change in understanding regarding the nature of the problem on the level of important states (Total) | 0 | | 1 | | 2 | | 3 | |
| 1 | 612 (50.8%) | 239 | 99.6% | 106 | 63.5% | 65 | 22.6% | 202 | 39.5% |
| 2 | 593 (49.2%) | 1 | 0.4% | 61 | 36.5% | 222 | 77.4% | 309 | 60.5% |
| **Total** | 1205 (100%) | **240** | 100% | **167** | 100% | **287** | 100% | **511** | 100% |

PROBLEM_UNDERSTAND_CHANGE_CAUSAL: 0 = not applicable (no change occurred), 1 = little or no causal impact, 2 = modest causal influence, 3 = significant causal influence
PROBLEM_UNDERSTAND_CHANGE (Change in Understanding regarding the Nature of the Problem on the Level of Important States): 1 = little or no change, 2 = significant change

**Table 5.4b**
Change in completeness of information about policy options on the level of important states

| | | OPTIONS_PROBLEM_CHANGE_CAUSAL (RC13) | | | | | | | |
|---|---|---|---|---|---|---|---|---|---|
| OPTIONS_PROBLEM_CHANGE (RC13) | Change in completeness of information about policy options in important nations | 0 | | 1 | | 2 | | 3 | |
| 1 | 612 (52.9%) | 224 | 99.5% | 96 | 64.9% | 113 | 36.7% | 179 | 37.7% |
| 2 | 544 (47.1%) | 1 | 0.5% | 52 | 35.1% | 195 | 63.3% | 296 | 62.3% |
| **Total** | **1156 (100%)** | **225** | **100%** | **148** | **100%** | **308** | **100%** | **475** | **100%** |

OPTIONS_PROBLEM_CHANGE_CAUSAL: 0 = not applicable (no change occurred), 1 = little or no causal impact, 2 = modest causal influence, 3 = significant causal influence
OPTIONS_PROBLEM_CHANGE (Change in Completeness of Information about Policy Options in Important Nations): 1 = little or no change, 2 = significant change

**Table 5.4c**
Change in understanding about the nature of the problem on the level of selected important states

| | | PROBLEM_UNDERSTAND_CHANGE_CAUSAL (RC 12) | | | | | | | | |
|---|---|---|---|---|---|---|---|---|---|---|
| IMPORTANT NATION | PROBLEM_ UNDERSTAND_ CHANGE (Change in understanding about the nature of the problem) (RC12) | Total of PROBLEM | 0 | | 1 | | 2 | | 3 | |
| Argentina | 1 = Little or no change | 61 | 31 | 50.8% | 0 | 0% | 0 | 0% | 30 | 49.2% |
| | 2 = Significant change | 10 | 0 | 0% | 3 | 30.0% | 2 | 20.0% | 5 | 50.0% |
| Australia | 1 = Little or no change | 67 | 32 | 47.8% | 3 | 4.5% | 2 | 3.0% | 30 | 44.7% |
| | 2 = Significant change | 17 | 0 | 0% | 3 | 17.6% | 5 | 29.4% | 9 | 52.9% |
| China | 1 = Little or no change | 6 | 1 | 16.7% | 3 | 50.0% | 1 | 16.7% | 1 | 16.7% |
| | 2 = Significant change | 12 | 0 | 0% | 2 | 16.7% | 5 | 41.7% | 5 | 41.7% |
| Denmark | 1 = Little or no change | 9 | 3 | 33.3% | 0 | 0% | 2 | 22.2% | 4 | 44.4% |
| | 2 = Significant change | 26 | 0 | 0% | 0 | 0% | 16 | 61.5% | 10 | 38.5% |
| European Union | 1 = Little or no change | 8 | 3 | 37.5% | 4 | 50.0% | 0 | 0% | 1 | 12.5% |
| | 2 = Significant change | 18 | 0 | 0% | 4 | 22.2% | 11 | 61.1% | 3 | 16.7% |

| | | | | | | | | | | |
|---|---|---|---|---|---|---|---|---|---|---|
| Germany | 1 = Little or no change | 13 | 2 | 15.4% | 3 | 23.1% | 5 | 38.5% | 3 | 23.1% |
| | 2 = Significant change | 52 | 0 | 0% | 4 | 7.7% | 27 | 51.9% | 21 | 40.4% |
| India | 1 = Little or no change | 9 | 1 | 11.1% | 6 | 66.7% | 1 | 11.1% | 1 | 11.1% |
| | 2 = Significant change | 11 | 0 | 0% | 2 | 18.2% | 5 | 45.5% | 4 | 36.4% |
| Japan | 1 = Little or no change | 19 | 0 | 0% | 12 | 63.2% | 6 | 31.6% | 1 | 5.3% |
| | 2 = Significant change | 29 | 0 | 0% | 3 | 10.3% | 17 | 58.6% | 9 | 31.0% |
| New Zealand | 1 = Little or no change | 58 | 32 | 55.2% | 0 | 0% | 2 | 3.4% | 24 | 41.4% |
| | 2 = Significant change | 15 | 0 | 0% | 3 | 20.0% | 3 | 20.0% | 9 | 60.0% |
| United Kingdom | 1 = Little or no change | 74 | 35 | 49.3% | 5 | 6.8% | 1 | 1.4% | 33 | 44.6% |
| | 2 = Significant change | 31 | 0 | 0% | 6 | 19.4% | 9 | 29.0% | 16 | 51.6% |
| United States | 1 = Little or no change | 94 | 38 | 40.4% | 12 | 12.8% | 10 | 10.6% | 34 | 36.2% |
| | 2 = Significant change | 67 | 1 | 1.5% | 12 | 17.9% | 22 | 32.8% | 32 | 47.8% |
| USSR/Russian Federation | 1 = Little or no change | 68 | 32 | 47.1% | 2 | 2.9% | 4 | 5.9% | 30 | 44.1% |
| | 2 = Significant change | 46 | 0 | 0% | 6 | 13.0% | 14 | 30.4% | 26 | 56.5% |

PROBLEM_UNDERSTAND_CHANGE_CAUSAL: 0 = not applicable (no change occurred), 1 = little or no causal impact, 2 = modest causal influence, 3 = significant causal influence

PROBLEM_UNDERSTAND_CHANGE (Change in Understanding about the Nature of the Problem): 1 = little or no change, 2 = significant change

**Table 5.4d**
Change in completeness of information about policy options on the level of selected important states

| | | OPTIONS_PROBLEM_CHANGE_CAUSAL (RC 13) | | | | | | | | |
|---|---|---|---|---|---|---|---|---|---|---|
| IMPORTANT NATION | OPTIONS_ PROBLEM_ CHANGE (Change in the completeness of information about policy options) | Total of PROBLEM | 0 | | 1 | | 2 | | 3 | |
| Argentina | 1 = Little or no change | 61 | 35 | 57.4% | 0 | 0% | 0 | 0% | 26 | 42.6% |
| | 2 = Significant change | 11 | 0 | 0% | 1 | 9.1% | 0 | 0% | 10 | 90.9% |
| Australia | 1 = Little or no change | 64 | 35 | 54.7% | 2 | 3.1% | 1 | 1.6% | 26 | 40.6% |
| | 2 = Significant change | 20 | 0 | 0% | 2 | 10.0% | 4 | 20.0% | 14 | 70.0% |
| China | 1 = Little or no change | 4 | 0 | 0% | 1 | 25.0% | 2 | 50.0% | 1 | 25.0% |
| | 2 = Significant change | 12 | 0 | 0% | 2 | 16.7% | 4 | 33.3% | 6 | 50.0% |
| Denmark | 1 = Little or no change | 18 | 0 | 0% | 3 | 16.7% | 12 | 66.7% | 3 | 16.7% |
| | 2 = Significant change | 14 | 0 | 0% | 0 | 0% | 7 | 50.0% | 7 | 50.0% |
| European Union | 1 = Little or no change | 5 | 0 | 0% | 3 | 60.0% | 1 | 20.0% | 1 | 20.0% |
| | 2 = Significant change | 15 | 0 | 0% | 1 | 6.7% | 2 | 13.3% | 12 | 80.0% |

| | | | | | | | | | | |
|---|---|---|---|---|---|---|---|---|---|---|
| Germany | 1 = Little or no change | 22 | 0 | 0% | 8 | 36.4% | 11 | 50.0% | 3 | 13.6% |
| | 2 = Significant change | 39 | 0 | 0% | 3 | 7.7% | 18 | 41.2% | 18 | 41.2% |
| India | 1 = Little or no change | 6 | 0 | 0% | 4 | 66.7% | 1 | 16.7% | 1 | 16.7% |
| | 2 = Significant change | 12 | 0 | 0% | 2 | 16.7% | 4 | 33.3% | 6 | 50.0% |
| Japan | 1 = Little or no change | 12 | 0 | 0% | 8 | 66.7% | 3 | 25.0% | 1 | 8.3% |
| | 2 = Significant change | 32 | 0 | 0% | 5 | 15.6% | 11 | 34.4% | 16 | 50.0% |
| New Zealand | 1 = Little or no change | 58 | 35 | 60.3% | 0 | 0% | 2 | 3.4% | 21 | 36.2% |
| | 2 = Significant change | 15 | 0 | 0% | 1 | 6.7% | 2 | 13.3% | 12 | 80.0% |
| United Kingdom | 1 = Little or no change | 68 | 36 | 52.9% | 6 | 8.8% | 0 | 0% | 26 | 38.2% |
| | 2 = Significant change | 33 | 0 | 0% | 8 | 24.2% | 10 | 30.3% | 15 | 45.5% |
| United States | 1 = Little or no change | 90 | 38 | 42.2% | 9 | 10.0% | 15 | 16.7% | 28 | 31.1% |
| | 2 = Significant change | 69 | 1 | 1.4% | 12 | 17.4% | 20 | 29.0% | 36 | 52.2% |
| USSR/Russian Federation | 1 = Little or no change | 73 | 35 | 47.9% | 0 | 0% | 12 | 16.4% | 26 | 35.6% |
| | 2 = Significant change | 34 | 0 | 0% | 4 | 11.8% | 9 | 26.5% | 21 | 61.8% |

OPTIONS_PROBLEM_CHANGE_CAUSAL: 0 = not applicable (no change occurred), 1 = little or no causal impact, 2 = modest causal influence, 3 = significant causal influence

OPTIONS_PROBLEM_CHANGE (Change in the Completeness of Information about Policy Options): 1 = little or no change, 2 = significant change limited to developing countries. Countries like the United States, Japan, Germany, the United Kingdom, and Russia have been able to improve their knowledge regarding the nature of problems and policy options for addressing them through collaboration in international environmental regimes. Even countries that exhibit high levels of economic development and possess well-developed scientific programs of their own benefit from programmatic activities involving assessment, monitoring, research, implementation review, and analyses of the adequacy of commitments. This is not to say that these countries always experience changes in available knowledge as a result of programmatic activities. But the evidence that regimes can contribute to learning on the part of policymakers operating in important states is striking.

## 5.5 Empirical Evidence—Knowledge and Problem Solving

The findings we have reported in the preceding section justify a degree of optimism regarding the role of institutions in the development of consensual knowledge. In some respects, these findings bolster cognitivist arguments regarding the effects of programmatic activities in producing consensual knowledge that can influence how states define their interests and determine their preferences. But the question remains whether the growth of consensual knowledge translates in turn into outcomes that are better in terms of problem solving. In theory, we should expect such a link to occur because enhanced knowledge should provide policymakers with a more sophisticated grasp of the nature of the problems under consideration and encourage them to select policy options that prove more effective in solving these problems.

Yet there is nothing automatic about such links. The application of improved knowledge requires leadership on the part of policymakers who use this knowledge to overcome the resistance of various interest groups and to mobilize support on the part of the attentive public. Even those who acknowledge in public the importance of confronting environmental problems can and often do drag their feet when it comes to implementation because the solutions to these problems conflict with the interests of well-organized groups in society. What is more, efforts to make use of improved knowledge may require adjusting the terms of the relevant regimes to extend programmatic activities, in order to include measures like financial assistance and technology transfers designed to encourage active engagement on the part of developing countries. Even willing parties frequently find that bringing about such adjustments is easier said than done.

To explore the links between improvements in knowledge and more effective outcomes, we developed two additional queries:

*Query 7* Are there discernible links between observed levels of understanding regarding the nature of the problem and changes in the status of the problem?

*Query 8* Are there discernible links between the completeness of information about the options available for dealing with the problem and changes in the status of the problem?

Table 5.5a contains data drawn from the IRD regarding the links between understanding of the problem and problem solving. We were able to examine this relationship empirically in a total of 188 records. Broadly speaking, the data indicate that there is a positive connection between knowledge and the status of environmental problems. One way to see this is to observe that the cells of the table toward the lower left are more heavily populated than the other cells in the table. Where the problem improved considerably, understanding of the problem was strongly or very strongly established in 37 cases. The comparable figure in cases where the problem improved slightly is 29. By contrast, there were only 3 cases in which the problem improved considerably in the face of a low established understanding and 7 cases of slight improvement under conditions of low understanding. By the same token, comparatively few cases resulted in the problem worsening despite the existence of a strongly established understanding. Thus, there are only 3 cases of considerable worsening and 2 cases of slight worsening where the understanding of the problem is very strongly established. Overall, then, the evidence drawn from the IRD does indicate that improvements in consensual knowledge are likely to prove helpful in efforts to solve environmental problems.

The links between information about policy options and changes in the status of problems are less clear (table 5.5b). Among the 176 cases

**Table 5.5a**
Links between the understanding about the nature of the problem and the state of an environmental problem

| | | PROBLEM_UNDERSTAND (RF 22) (Understanding about the nature of the problem) | | | | | | | |
|---|---|---|---|---|---|---|---|---|---|
| PROBLEM_ CHANGE (RC 11) (Dimensions of change in the state of a problem) | Total | 1 | | 2 | | 3 | | 4 | |
| 1 | 20 (10.6%) | 3 | 7.1% | 7 | 8.6% | 6 | 14.0% | 4 | 18.2% |
| 2 | 35 (18.6%) | 2 | 4.8% | 14 | 17.3% | 12 | 27.9% | 7 | 31.8% |
| 3 | 35 (18.6%) | 10 | 23.8% | 21 | 26.0% | 3 | 7.0% | 1 | 4.5% |
| 4 | 50 (26.6%) | 13 | 31.0% | 16 | 19.8% | 14 | 32.6% | 7 | 31.8% |
| 5 | 48 (25.5%) | 14 | 33.3% | 23 | 28.4% | 8 | 18.6% | 3 | 13.6% |
| **Total** | **188 (99.9%)*** | **42** | **100%** | **81** | **100.1%** | **43** | **100.1%** | **22** | **99.9%** |

PROBLEM_UNDERSTAND (Understanding about the Nature of the Problem): 1 = very strongly established understanding, 2 = strongly established understanding, 3 = partially established understanding, 4 = low established understanding

PROBLEM_CHANGE (Dimensions of Change in the State of a Problem): 1 = the problem worsened considerably, 2 = the problem worsened slightly, 3 = the problem stayed the same, 4 = the problem improved slightly, 5 = the problem improved considerably

**Table 5.5b**
Links between the completeness of information about policy options and the state of an environmental problem

| | | INFORMATION_OPTIONS_PROBLEM (RF 23) (Completeness of information about policy options) | | | | | |
|---|---|---|---|---|---|---|---|
| PROBLEM_ CHANGE (RC 11) (Dimensions of change in the state of a problem) | Total | 1 | | 2 | | 3 | |
| 1 | 17 (9.7%) | 1 | 2.7% | 6 | 5.8% | 10 | 27.8% |
| 2 | 35 (19.9%) | 7 | 18.9% | 19 | 18.4% | 9 | 25.0% |
| 3 | 27 (15.3%) | 11 | 29.7% | 10 | 9.7% | 6 | 16.7% |
| 4 | 50 (28.4%) | 3 | 8.1% | 40 | 38.8% | 7 | 19.4% |
| 5 | 47 (26.7%) | 15 | 40.5% | 28 | 27.2% | 4 | 11.1% |
| Total | 176 (100%) | 37 | 99.9% | 103 | 99.9% | 36 | 100% |

INFORMATION_OPTIONS_PROBLEM (Completeness of Information about Policy Options): 1 = very high completeness, 2 = medium completeness, 3 = low completeness
PROBLEM_CHANGE (Dimensions of Change in the State of a Problem): 1 = the problem worsened considerably, 2 = the problem worsened slightly, 3 = the problem stayed the same, 4 = the problem improved slightly, 5 = the problem improved considerably

for which there are data in the IRD, 15 fall into the cell featuring a high level of completeness of information and considerable problem improvement. Another 28 belong to the cell combining medium completeness and considerable problem improvement. On the other hand, medium completeness is associated with slight problem improvement in 40 cases; the problem stayed the same in 11 cases characterized by high completeness, and there are 19 cases in which medium completeness of information is associated with a slight worsening of the problem. Thus, it appears that there is some connection between information about policy options and considerable improvement in the status of problems. But we cannot conclude that there is any generic connection between the completeness of information about policy options and problem solving in this empirical domain.

Clearly, there are links between the evolution of consensual knowledge and the state of the environment. But these links are not as clear and straightforward as the theory would lead one to expect. We should note as well that the benchmark used in this chapter to assess environmental problem solving is less demanding than the idea of a collective optimum advanced in some of the literature on the effectiveness of regimes (Helm and Sprinz 1999). For pragmatic reasons, we have focused on relative improvement or worsening in the state of the problem over time. This avoids the problem of operationalizing the idea of the collective optimum on a case-by-case basis as an element in the precoding agreement or relying on coders to make their own judgments about this matter. We acknowledge that some of the links between knowledge and environmental problem solving might have been less significant if we had made use of the idea of the collective optimum as a benchmark. Nevertheless, our procedure has much to recommend it, at least among those seeking to make good use of the IRD as a research tool. The idea of the collective optimum is elusive. Given the problems they encountered in handling other matters, it seems clear that the coders would have had a hard time dealing with the idea of the collective optimum.

## 5.6 Conclusion: The Power of Knowledge

The empirical evidence presented in this chapter offers some support for theoretical arguments about the role of knowledge in addressing collective-action problems dealing with environmental affairs. But an examination of the data also makes it clear that the growth of knowledge resulting from a regime's programmatic activities does not translate automatically into greater success in the realm of problem solving. Many other factors also come into play in determining outcomes arising from human-environment interactions both in local situations (e.g., a small-scale fishery) and in global matters (e.g., climate change).

While there is evidence that the growth of knowledge arising from programmatic activities enhances understanding of the nature of the problem and completeness of information about policy options, the links relating to completeness are weaker than those relating to understanding of the nature of the problem. This may be due, in part, to the fact that it is harder to forecast future consequences arising from specific policy choices than to reduce uncertainty with regard to the biophysical and socioeconomic forces involved in existing problems. It may also testify to the fact that policymakers are often committed to the use of certain policy instruments, regardless of the nature of the problem or efforts to foresee how those instruments will work in specific situations.

All this suggests the need for caution in thinking about the links between programmatic activities and environmental problem solving, as well as the need to become more proactive in the production and dissemination of knowledge about the range of policy options available to deal with specific problems. But it does not lead to the conclusion that the links between knowledge and problem solving are weak and unworthy of scholarly attention. Even though the data do not provide decisive evidence in support of the cognitivist perspective on the links between knowledge and problem solving, our analysis indicates clearly that arguments regarding the role of knowledge are worthy of more sustained investigation on the part of those seeking to understand why some regimes are more effective than others in addressing the problems that lead actors to create them. It is fair to say in concluding this chapter that the IRD is as useful in shaping the agenda for further research as it is in "testing" hypotheses drawn from the existing literature on the effectiveness of environmental regimes.

# 6

# Conclusion: Key Findings and Future Directions

## 6.1 Introduction

We are aware that some analysts may regard this book as untimely, for both political and theoretical reasons. One political criticism, exemplified in the work of Bjørn Lomborg, asserts that we are making good progress in addressing environmental as well as social problems, and that we do not need to expend a great deal of time and energy on the arduous process of creating and implementing new institutional arrangements to deal with problems of this sort. As Lomborg puts it in his book *The Skeptical Environmentalist* (2001, 348), "Conditions in the world are not getting worse and worse.... We have more leisure time, greater security and fewer accidents, better education, more amenities, higher incomes, fewer starving, more food, and healthier and longer lives. There is no ecological catastrophe looming around the corner to punish us." Those who subscribe to this sunny view are apt to regard elaborate efforts to analyze the effectiveness of international environmental regimes as a waste of time.

Others will consider our effort as untimely in political terms for exactly the opposite reason. They do see a catastrophe looming around the corner, and they regard international regimes as wholly inadequate to provide the tools of governance needed to cope with these environmental problems. Gus Speth articulates this view with particular clarity. Reviewing the rapid growth in the number of environmental conventions and treaties over the last thirty years, he observes in his book *Red Sky at Morning* (2004, xi) that

> superficially, this outpouring of international legislation was similar to the burst of domestic environmental legislation of the 1970s. But, unfortunately, the similarity stopped there. Whereas our national legislation was successful in curbing many environmental abuses domestically, efforts to protect the global environment have largely failed in the sense that the trends in environmental deterioration have not improved and that more of the same will not get us where we want to be in time to head off an era of unprecedented environmental decline. Big trouble is coming down the pike—and coming fast indeed.

From this perspective, the development of issue-specific regimes to handle discrete environmental problems is, at best, an exercise in applying Band-Aids to festering wounds.

From a theoretical perspective, some readers are likely to criticize our effort to separate environmental politics into discrete issues and issue areas. The IRD and the analysis presented in this book build on the premise that it is useful to differentiate among distinct issue areas in studying the roles that institutions play in international society. In fact, we go a step further, breaking regimes into components associated with their own constitutive agreements and stages separated by temporal watersheds and using the individual regime element as our basic unit of analysis. This approach runs counter to the view emphasized in some recent, largely constructivist analyses that individual regimes are deeply embedded in the broader setting of international society and that individual regimes interact extensively with one another as well.

Those who subscribe to this constructivist perspective are likely to question the selection of the individual regime—much less the regime element—as the basic unit of analysis for efforts to improve our understanding of the roles that institutions play in international society. They understand that more or less identical regimes can succeed in one setting and fail in another due to the impact of the social contexts in which they operate. They point as well to the importance of institutional interplay as a determinant of the performance of issue-specific regimes. In addition, constructivists will conclude that the IRD Data Protocol omits a number of important ideas and misses factors that affect the performance of regimes in international society. Sophisticated views regarding the influence of discourses exemplify this critique.

Still others will regard the IRD as too state-centric due to the fact that the regimes included in the database all involve intergovernmental agreements in which states are the principal members. This cuts against the current fashion emphasizing the roles of civil society organizations and nonstate actors and drawing attention to emerging forms of private governance, including codes of conduct and various types of corporate social responsibility.

We respect these views and even find some of the current intellectual fashions appealing. But we do not agree that they marginalize or undermine the sorts of analyses made possible by using the International Regimes Database. A review of the regimes included in the IRD demonstrates that many international environmental problems are serious, even when they do not constitute an immediate threat to the planet's life support systems. Whatever our current concerns about global problems, such as climate change and the loss of biological diversity, the IRD shows clearly not only that many problems are solved or substantially ameliorated but also that regimes play important roles in achieving these results. Although institutional interplay and the social as well as ideational embeddedness of regimes undoubtedly are matters of growing concern, this development does not preclude the analysis of a wide range of other matters, with regard to which it makes good sense to focus on the individual regime or even the regime element as the basic unit of analysis. The activities of nonstate actors are on the rise at the international level, and the growth of what have become known as Type II partnerships is clearly an interesting development. Nonetheless, states remain central to efforts to improve governance in international society. More often than not, the constitutive provisions that define the institutional character of international regimes are spelled out in international conventions, treaties, or other constitutive agreements.

In this concluding chapter, we respond first to the political criticisms by showing in Section 2 that issue-area specific regimes often do matter and sometimes play a decisive role in ameliorating environmental problems. In Section 3, we address the theoretical and methodological criticisms by clarifying some limits of the database but also by exploring the unused potential of this information utility more extensively. We take the position that research based on the IRD can address a range of issues in a manner that allows us to go well beyond previous analyses based on discrete case studies. In this sense, the IRD provides a point of departure for major advances in our knowledge of international environmental governance. Section 4 focuses on future directions; it offers some brief comments regarding next steps and encourages others to take advantage of the IRD in their efforts to strengthen and expand regime analysis.

## 6.2 Key Findings

The substantive findings presented in this book deal with the effectiveness of international regimes or, in other words, the extent to which the operation of these institutions can account for variance in the outcomes of interactions among actors in international society. We have approached the issue of effectiveness at several levels, asking about the roles that regimes play in eliciting compliance with rules and regulations, in attaining the goals spelled out in constitutive documents, and in solving the problems that lead to their creation in the first place. In each case, the basic conclusion is the same. Regimes matter in the sense that they are significant determinants of collective outcomes. Of course, effectiveness is a variable. Regimes are not the only forces that make a difference in this context, and the proportion of the variance attributable to the operation of regimes differs from case to case. But this in no way detracts from the proposition that regimes deserve sustained consideration as one important determinant of collective outcomes at the international level.

Our initial conclusion, then, is that it is high time to set aside sterile debates about regime effectiveness couched in general terms and to move on to more focused analyses of the extent to which regimes matter, the ways in which they matter, and the circumstances affecting the proportion of the variance in collective outcomes we can attribute to the operation of regimes. Our empirical analyses based on the IRD and presented in the preceding chapters have a good deal to tell us about these issues. In this section, we summarize and comment on our principal conclusions under four broad headings: the big picture, theoretical implications, analytic tensions, and generalizability.

### 6.2.1 The Big Picture

Environmental regimes are intended ordinarily to solve problems of governance ranging from global concerns like the depletion of stratospheric ozone to spatially limited concerns like protecting the Great Lakes of North America as well as from broad issues like managing a range of human uses of marine resources to narrower issues like protecting polar bears or fur seals. Yet a simple analogy to the functions of governments in domestic politics does not offer much help in understanding when, how, and why international regimes work. In fact, the analyses we have presented in the substantive chapters of this book make it clear that the approach to governance embedded in international regimes differs fundamentally from our conventional view of the functions of governments in the ordinary sense of the term. To understand the significance of international regimes, therefore, we need to develop an analytic framework and a mode of reasoning that is designed explicitly to deal with the performance of these arrangements.

Consider, in this connection, three observations regarding the distinctive roles that regimes play drawn from the substantive chapters of this book. Almost 54 percent of our cases rely in practice on the use of consensus rules to arrive at collective choices; unanimity is the decision rule of choice in another 20 percent of the cases. This is widely regarded—especially among those who think in terms of domestic systems—as a serious drawback. The general expectation is that governance systems relying on decision rules of this type will produce decisions conforming to the preferences of the least ambitious member of the group and that this will often amount to sticking with the status quo (Miles et al. 2002). But the evidence from the IRD does not support this conclusion. Problem improvement occurs in half the cases where regimes rely on consensus rules and in 37 percent of the cases making use of unanimity rules. What is more, coders judged that regimes make a significant or very strong difference in 70 percent of the cases where problems improve considerably and consensus rules are in place. The comparable figure for regimes featuring unanimity rules is over 50 percent. The inference we draw from this evidence is that it does not help to rely on simple analogies to domestic practices in seeking to understand how international regimes work. Consensus building is an elastic process. Particularly in cases where the preferences of individual actors are not set in stone at the outset and where uncertainty plays a prominent role, the process of consensus building can generate important insights about the relative merits of different ways to approach problems and eventuate in results that no one could have foreseen—much less predicted—at the outset.

We have a similar story to tell with regard to compliance. A simple analogy to domestic politics suggests that compliance with inconvenient rules is highly unlikely in the absence of hierarchy (Downs, Rocke, and Barsoom 1996). But our data indicate that the behavior of members of international regimes meets fully or exceeds regime requirements over 60 percent of the time. Moreover, our coders judged the causal role of the regime in eliciting compliance to be large in almost two-thirds of the cases in which behavior meets or exceeds requirements. At the same time, an analysis of the records included in the IRD does not provide overwhelming support for the approaches to compliance generally associated with the concepts of enforcement and management (Chayes and Chayes 1995). Sanctions of one sort or another do make a difference. But reporting procedures and verification do not loom large, suggesting that transparency may not be as potent as many individual case studies have suggested. Our evidence suggests as well that some of the factors emphasized in the management approach (e.g., capacity building) do not play a powerful role in raising levels of conformance with the rules and regulations of regimes. What do these findings mean? We have found that two other forces often overlooked in analyses of compliance—legalization and legitimacy—account for a sizable proportion of conformance on the part of regime members.

Most of those who study regimes treat rules—in the form of requirements and prohibitions—as the core of these institutional arrangements, a view that leads naturally to a focus on implementation and compliance as the central concern in efforts to understand the effectiveness of regimes (Krasner 1983a). Our findings do not deny the importance of rules. But they do point to the critical role of programmatic activities as determinants of the effectiveness of regimes (Young 1999b, chap. 2). What this means is that there is often considerable room for (re)adjustments in the framing of environmental problems, in the compilation of information about feasible response strategies and their probable results, and even in assessments of the consequences arising from specific strategies for individual regime members. In extreme cases, programmatic activities can play a role in altering the discourses in terms of which problems are discussed. Something of this sort seems to have happened in the case of stratospheric ozone (Litfin 1994). But even short of the emergence of new discourses, regimes can affect outcomes by influencing problem definitions and structuring the content of the set of problem-solving strategies considered. We conclude, in this connection, that models that assume these factors away at the outset (e.g., many game-theoretic and microeconomic models) run the risk of missing some important processes through which regimes can make a difference in efforts to come to term with environmental problems.

### 6.2.2 Theoretical Implications

Let us focus a little more intensively on these findings in order to tease out the implications of our analysis regarding the behavior of actors in international society. The dominance of neorealism and neoliberalism in the field of international relations during the last twenty-five to thirty years has had the effect of narrowing and often polarizing our thinking about the sources of actor behavior (Waltz 1979; Keohane 1984). Thus, we tend to treat states as unitary actors that make choices either because powerful actors coerce them into selecting certain alternatives or because the choices they make arise from incentives and reflect straightforward utilitarian calculations. We do not deny the role of power and interests as determinants of behavior in international society. But the evidence presented in the preceding chapters suggests that it would be a serious mistake to assume that neorealism or neoliberalism can account for everything that happens at the international level.

What should we make of the conclusions reported in chapter 4 regarding the generally weak links between the operation of compliance mechanisms and goal attainment and problem solving? An obvious inference is that the behavior of the relevant actors is not always driven by incentives that can be represented in conventional utilitarian or benefit-cost calculations. But this observation begs the question. If incentives are not the key to understanding the behavior of those subject to a regime's rules, how can we account for observed outcomes in this realm? The data presented in chapter 3 provide some help in answering this question. Both legalization in the forms of juridification and internalization and the development of legitimacy make a substantial difference. One inference to be drawn from this line of analysis is that the rule of law matters (Allott 1999), even in a setting where procedures designed to produce authoritative interpretations are underdeveloped and meaningful sanctions are seldom imposed. It may be that this is primarily a tribute to the effectiveness of fears regarding the breakdown of social order as a force impelling actors to take the rule of law seriously. But in any case, there is something going on here that is not easily understood in terms of the influence of power and interests.

Similar comments are in order regarding the findings we report in chapter 5 concerning programmatic activities. At the most fundamental level, our findings suggest that it is not helpful to assume that well-informed actors with predetermined interests or preferences enter into agreements about environmental problems that can be understand as issue-specific social contracts. In fact, many regimes deal with issues that are poorly understood; individual members often lack both the factual information and the theoretical understanding that would be needed to forecast the probable impacts of the operation of regimes on their interests. There are even cases in which members of the same regime define the problem differently (e.g., decarbonization of postindustrial societies versus controlling concentrations of carbon dioxide in the earth's atmosphere in the case of climate change). What makes for an effective regime in a setting of this sort? While there is no simple answer to this question, it is clear to us that regimes can make a difference by shaping the ways participants frame issues, define problem-solving strategies, and evaluate results. Interestingly, individual participants may be poorly equipped to forecast the probable results of participation at the time they have to make a choice concerning whether or not to join. But the evidence from the IRD does not indicate that this is a serious barrier to the formation and implementation of environmental regimes in any overall sense.

More generally, we interpret the findings reported in the substantive chapters of this book as at least a partial challenge to the highly suggestive and popular distinction between the logic of consequences and the logic of appropriateness (March and Olsen 1998). Roughly speaking, this distinction suggests that the behavior of actors in international society is driven by some combination of utilitarian calculations and normative commitments. That is, actors make the choices they do either because they represent the best alternatives (i.e., the options selected have the highest benefit-cost ratios) or because they constitute the right alternatives (i.e., the options chosen conform best to guiding normative principles).

There is no need to deny the role of consequences and appropriateness to appreciate that they do not offer—singly or in combination—a general theory of the behavior of the actors in international society. To be more specific, our findings point to the significance of three sources of behavior that do not fit easily into mainstream accounts. Actors often choose options that conform to the precepts of a knowledge system or a discourse that has come to dominate thinking about a particular issue (Litfin 1994). Much of the current debate about responses to climate change, for instance, is really a confrontation between dueling discourses; communication phrased in these terms can produce a significant impact on behavior (Müller 1994; Risse 2000). There are also clear indications that perceptions of legitimacy matter. Whether an issue arises at the local level or the global level, stakeholders who believe that the rules of the game have emerged from a process that is fair or equitable are more likely to abide by them on a day-to-day basis than stakeholders who feel no sense of ownership of the process (Franck 1990). Beyond this, habits or standard operating procedures play a significant role. The secret of success in most effective social practices lies in the fact that following the rules becomes second nature for most participants. They conform to the rules on a day-to-day basis without giving it a second thought; some actually absorb the basic features of social practices into the ways they think about their identity (Checkel 2001).

The implications of this line of reasoning are far-reaching. We do not wish to take sides in the methodological battles among those who argue for the primacy of power or interests or knowledge as determinants of collective outcomes at the international level. But the findings we have reported in this book do reinforce the proposition that there is a need to move beyond simple assumptions that treat the members of international regimes as unitary actors making rational choices in order to understand the roles institutions play in international society. The issues at stake here are much the same as those arising in conjunction with the growing influence of behavioral economics in the field of microeconomics. There is no need to abandon a decision-theoretic approach to understanding problems of governance in international society. But we need models that are more realistic than those underlying much theorizing about collective-action problems in this social setting (Kahneman 2003). Moving in this direction will incur some costs measured in terms of the tractability of our analytic constructs. But these costs will be offset by substantial gains in terms of the relevance of our findings to efforts to solve real-world problems. In general, our findings suggest the results will also be compatible with the development of a body of knowledge that accords a substantial—though by no means determinative—role to institutions as factors that account for a significant fraction of the variance in collective outcomes in international society.

### 6.2.3 Analytic Tensions

Alert readers will have noticed that some of the findings we have reported in the previous chapters seem, at least on the surface, to be at odds with one another. What can we say about these puzzling conclusions? Closer analysis often suggests that the findings in question are not strictly contradictory. But a discussion of these tensions yields several significant insights.

Queries about the impact of compliance mechanisms on goal attainment and problem solving led us to conclude that the management approach to compliance (Chayes and Chayes 1995)—in contrast to the enforcement approach—is dominant in international environmental regimes. To be specific, coders supplying information on individual regimes concluded that the management approach dominates efforts to elicit compliance in approximately 90 percent of the cases. On the other hand, our analysis of factors that play a role in encouraging regime members to comply with the rules suggests that approaches relying on various forms of capacity building are not particularly effective.

How is this possible? We are not in a position to provide a decisive answer to this question. Even so, several preliminary observations are in order. Capacity building per se is only one element in the combination of factors commonly lumped together under the heading of the management approach. While the idea of enforcement refers specifically to compliance mechanisms that affect behavior by driving up the costs of noncompliance or (less often) increasing the benefits of compliance, the idea of management encompasses a variety of behavioral mechanisms. What ties these mechanisms together is an emphasis on sources of behavior that do not emphasize the logic of consequences. Mechanisms featuring internalization as well as the cultivation of a sense of ownership on the part of stakeholders, for instance, belong to this cluster of driving forces. Thus, a number of the mechanisms discussed in chapter 3 belong to the broad and sometimes ill-defined category of management, as opposed to enforcement. On this account, capacity building is one mechanism that is relevant or useful in some situations. The Montreal Protocol Multilateral Fund, for example, fits this description as a mechanism for encouraging developing countries (e.g., China, India) to join the regime dealing with the depletion of stratospheric ozone. But this does not imply that the management approach to compliance always involves a strong emphasis on capacity building.

Although less prominent in the data, somewhat similar comments are in order regarding the role of compliance mechanisms in the operation of effective regimes. Whereas chapter 3 suggests that strong verification procedures and horizontal compliance mechanisms play a role in eliciting compliance, chapter 4 reports that the data in the IRD do not point to any strong links between specific compliance mechanisms and goal attainment or problem solving. In this connection, it is important to distinguish between the variables covering procedures dealing with reporting and implementation review (Forms RA 45 and RA 46) and the variable focusing on formal (as opposed to de facto) compliance mechanisms (RA 47). Chapter 3 directs attention to Forms RA 45 and RA 46 and concludes that strong forms of verification make a difference in eliciting compliance. Chapter 4, by contrast, focuses on Form RA 47 and concludes that regimes rely on a small subset of the range of potential compliance mechanisms and that the links between the formal mechanisms used and success in attaining goals or solving problems are relatively weak.

Even though they do not contradict one another explicitly, the juxtaposition of these findings is puzzling. What seems to be going on here is a process involving compliance mechanisms that are significant in behavioral terms but that generally do not show up in the formal constitutive provisions of regimes. Factors like legalization and a sense of legitimacy loom large as determinants of compliant behavior in connection with many regimes. But these factors are seldom referred to in the constitutive agreements on which regimes are founded. Rather, factors of this sort emerge and become influential as aspects of the processes through which regimes evolve into social practices. These observations indicate, once again, the need to move beyond conventional assumptions about the behavior of actors in international society in order to understand the prospects for achieving success in governance without government in this social setting.

### 6.2.4 Generalizability

For the most part, common interests loom large relative to distributive concerns in successful environmental regimes. Some regimes (e.g., the regime dealing with ozone depletion) have encountered a need to address distributive issues in order to bring major developing countries into the system. In other cases (e.g., climate change, loss of biological diversity, desertification), the need to come to terms with distributive issues continues to limit success with regard to goal attainment or problem solving. In our judgment, a favorable balance of common and individualistic interests is critical to the development of a constituency that subscribes to an integrated interpretation of the nature of the problem to be solved, and is also crucial to the achievement of success in collective decision making in situations featuring the use of consensus rules.

Whether the existence of these conditions owes more to the character of the problems to be solved—various environmental problems in the case of the IRD—or to the effects of social construction is an interesting and important question. Many of the regimes included in the IRD are properly understood as arrangements designed to overcome market failures, such as the impacts of externalities or the consequences of the free-rider problem. Often, this gives rise to an emphasis on programmatic activities (e.g., the production of new knowledge regarding the impacts of various pollutants) intended to improve the ability of actors to design and implement policies that are well matched to the nature of the problems at hand. This emphasis tends to direct attention to environmental regimes whose membership and spatial domains are oriented toward the advanced industrial nations. Among other things, this is probably part of the explanation for the relative lack of emphasis on measures aimed at capacity building that we commented on earlier in this chapter.

This tendency to direct attention toward the developed world also produces a bias in favor of regimes created to cope with problems arising from rapid and prolonged industrialization. Thus, transboundary pollution affecting airsheds, international rivers basins, and regional seas as well as other externalities like the export of hazardous wastes are all negative side effects of the economic systems that are common to the members of the OECD. Both public concern about these issues and the willingness of policymakers to tackle them have grown substantially over the last two to three decades.

Still, environmental problems are growing rapidly among the countries of the developing world; many regimes (e.g., the ozone regime) that started out as initiatives of advanced industrial counties have expanded to include a wide range of developing countries (Hoffmann 2005). In our view, the next phase of research on the effectiveness of environmental regimes should feature a concerted effort to include a greater number of North-South arrangements. The availability of experts who are qualified to code such regimes will be a limiting factor, at least at the outset. But it should be possible to overcome this problem as North-South issues become more prominent and the number of analysts concerned with these issues grows.

The emergence of issues that have a strong North-South component (e.g., climate change and the loss of biological diversity) ensures that problems of fairness or equity will loom larger in the formation and implementation of environmental regimes during the foreseeable future. But these distributional concerns are by no means limited to the North-South context. Policymakers in many industrialized countries are increasingly sensitive to the economic burden associated with effective efforts to address large-scale environmental issues. As a result, we can expect a growing emphasis on the need to find ways to avoid the gridlock that can accompany an intense concern about distributive issues. We expect this will lead to an enhanced effort to develop institutional mechanisms capable of addressing issues of equity and a corresponding emphasis on the effectiveness of such mechanisms on the part of analysts studying the consequences of international environmental regimes.

## 6.3 Uses of the International Regimes Database

Our project has methodological as well as substantive goals. Building the IRD has been an experiment in the development of methods to facilitate scientific research in an area where individual variables are hard to measure, clusters of variables are often interactive, and universes of cases are typically small. Having reflected on the substantive implications of our findings, therefore, we turn in this section to a discussion of the methodological implications of our work. In this connection, we start with some observations about the limitations of the IRD and then move on to a discussion of unused potential and opportunities to use the database for research extending well beyond the scope of this book.

### 6.3.1 Limitations

Our quantitative analyses show beyond any reasonable doubt that international environmental regimes do make a difference; they regularly help to solve or ameliorate problems arising from human-environment interactions. What is more, institutional design plays a significant role in bringing about these positive results. Although these findings seem to us quite significant, we are well aware that the creation and operation of international regimes cannot explain all the variance regarding human-environment interactions we see in the real world. This leads us to comment on three limitations that need to be kept in mind by those interested in using the IRD to shed light on real-world situations.

Currently, the data in the International Regimes Database pertain exclusively to environmental regimes, even though we designed the IRD Data Protocol to allow for the addition of data on other types of regimes. It follows that we need to be extremely cautious in applying our findings to arrangements dealing with other issue areas or to international regimes in general. Still, we believe there is no reason to assume a priori that environmental politics differ dramatically from the rest of politics.

Disaggregating the universe of international regimes into discrete elements in order to facilitate quantitative comparisons is also a bold step. Proper comparisons build on the assumption that the individual units to be compared are independent of one another. To the extent that regimes interact with each other or are embedded in a common overarching normative framework, this assumption is compromised. In one sense, the approach we adopted in designing the IRD aggravates this problem. The use of the regime element as our basic unit of analysis makes the assumption of independence among individual units even more problematic. In a factual sense, it is undoubtedly wrong to assume that a regime following a watershed is fully independent of the arrangement as it existed before the watershed. It is also a stretch to assume, for instance, that the $SO_2$ protocol of the regime for long-range transboundary air pollution is independent of the $NO_X$ protocol.

Of course, this problem is general in nature; it applies to almost all comparative analyses in political science or, for that matter, in the social sciences more generally. Comparisons across national political systems, for instance, run into the same predicament. In a globalized world featuring a panoply of material interdependencies and opportunities to learn from the experiences of others, the independence of nations and national political systems vis-à-vis one another has vanished (if it was ever there). Under the circumstances, the best we can do is to be alert to the occurrence of such connections and to correct for them in interpreting data included in the IRD. The embeddedness of individual regimes in a larger normative system and the interplay among specific international regimes are among the most important of these linkages in our field of study. We regard this problem as an important one. But we do not believe that the occurrence of interdependencies among our units of analysis invalidates the strategy we have developed to pursue understanding of matters like the determinants of the effectiveness of international regimes.

A third problem involves theoretical limitations of the IRD Data Protocol. The emerging emphasis on issues pertaining to embedded norms and interplay among individual regimes goes hand in hand with the rise of interest in issues that are especially relevant to constructivist theorizing about international institutions. When we constructed the IRD Data Protocol, constructivist thinking in international relations was not well developed. Since then, many theoretically interesting ideas calling for rigorous empirical testing have come into focus. Hypotheses relating to the interplay of international regimes are good examples. So are questions regarding the legalization of international norms and the role of bodies created to settle disputes. Similar comments apply to the role of communicative actions and the effects of epistemic communities and larger systems of ideas or discourses. Because the articulation of these ideas in a precise form is a recent development, they did not find their way into the IRD.

This is an unavoidable limitation of any long-term empirical project. It would have been presumptuous to include unsystematic speculations about future developments in international relations theory in the IRD Data Protocol. Our experience makes it clear as well that it is not feasible to go back to the same coders repeatedly with requests that they update their responses to include recent developments or to provide information regarding new variables. In the end, we concluded that the best strategy is to treat the database as a completed product and to address factors not included in the database by other means.

Yet this does not mean that the database has nothing to say about newly emerging issues in the field of regime analysis. The broad scope of the IRD Data Protocol suggests the value of a search for proxy measures. Although the database does not address some new issues explicitly, it may contain information about matters that can serve as surrogates for new issues. We have taken some steps in this direction in the substantive chapters of this book. In chapter 3, for instance, we treat the level of autonomy of secretariats and other organizational arrangements as a proxy for the existence of autonomous dispute-settlement procedures. In this way, we have sought to explore an important idea in the current debate about legalization via measures framed and included in the IRD Data Protocol before this debate arose.

A second strategy features combining data included in the IRD with additional data on measures not included in the database. If analysts wish to investigate the relationship between arguing and bargaining in the conventional sense during processes of regime formation, for example, they could combine new data on the balance between arguing and bargaining with data from the IRD dealing with various aspects of regime formation (e.g., information concerning the cognitive setting) or regime effectiveness (e.g., information on the impacts of regimes in their targeted issue areas). Such combinations are attractive because they reduce the level of effort required to carry out important studies regarding a wide range of issues.

We note as well that the same general problem arises in conjunction with other efforts to create databases (e.g., the world values survey). The typical response is to treat each round of database building as having two components. One features the repetition of earlier questions to analyze developments occurring over time. The second component involves the addition of new questions designed to address issues that have emerged during the period since the construction of the database and that deserve inclusion in subsequent rounds of database building. In the case of the IRD, we believe the first of these procedures will be difficult or infeasible due to problems with coding and coders. But future work on international regimes could well follow the second path, adding new information on matters that have come into focus since we created the IRD Data Protocol during the 1990s.

### 6.3.2 Unused Potential

Clearly, then, the IRD has limitations; the strategies discussed above cannot avoid or eliminate these constraints completely. But at the same time, the potential of the IRD as a tool for studying international institutions reaches far beyond the findings we have reported in this book. Chapters 3 to 5 illustrate the use of the database to expand our knowledge of specific issues relating to effectiveness. These issues are certainly important. But they do not begin to encompass the potential uses of the IRD. In principle, analysts can use each of the 136 variables included in the database as either a dependent variable or an independent variable. For each independent variable, there are numerous conceivable effects; each dependent variable can have numerous possible causes. We can amplify this general point by looking briefly at each of the four sections of the IRD and highlighting examples of studies made possible by the development of this research tool.

None of the empirical chapters of this book focuses on regime formation. Yet the IRD contains data for fifty-seven variables relevant to the analysis of regime formation. Information about the development of regimes from agenda setting through negotiation and on to implementation is accompanied by information about the most important causes of regime formation identified in regime analysis. Among the possible causes of regime formation included in the IRD are the features of the problem to be solved, the political setting, the economic setting, the cognitive setting, and the role of important actors.

There are also opportunities to explore ideas developed in the literature on the rational design of institutions (e.g., the effects of problem structure or the cognitive setting) in accounting for choices relating to the selection of regime attributes (Koremenos, Lipson, and Snidal 2001). An unexplored but potentially fruitful line of research would feature an effort to identify links between the political constellations existing prior to regime formation and regime consequences or even regime dynamics (Gehring 1994). All these inquiries—and many others—are feasible with the help of the IRD, although we have not tackled them in this book.

The IRD contains information on a wide range of regime attributes. In designing the database, we took the claims of institutionalists seriously and cast a wide net. The data in this section of the database encompass norms and ideas involving underlying visions and theories as well as organizational features (e.g., rules dealing with membership and funding mechanisms). Overall, this section includes fifty-one variables, opening the window for developing a broad research program comparable to the effort in comparative politics to evaluate the impact of specific features of political systems on policy outputs (Weaver and Rockman 1993). The most obvious examples involve studies of links between regime attributes and regime consequences and dynamics. Thus, analysts can ask any number of questions about the extent to which various features of regimes play a role in increasing or decreasing the effectiveness of environmental regimes. It is possible as well to launch similar inquiries about links between regime attributes and the formation and dynamics of these institutional arrangements. Questions about what types of regimes form and under what conditions, for example, are both important and relatively understudied.

The third and the fourth sections of the IRD focus on regime consequences and regime dynamics. The IRD contains data covering regime outputs, outcomes, impacts, and broader consequences as well as data relating to processes of institutional change. As we observed earlier, the IRD allows for systematic studies of the effects of processes of regime formation on the success of regimes in attaining goals or solving problems. But beyond this, we can ask whether regimes created with ease and without serious resistance generate a stronger compliance pull or produce better results than other arrangements. More generally, the IRD may make it possible to enhance our understanding of regime effectiveness beyond that achieved in the most sophisticated prior research (Miles et al. 2002).

Finally, there are many cases in which it will be worthwhile to study relationships between variables located in a single section of the IRD. A prominent example that we have chosen as a focus of analysis in this book involves links between compliance with rules and regime effectiveness. Are compliance and effectiveness really so different that high rates of compliance cannot be construed as an indicator of effectiveness? Should the relationship between these two variables turn out to be stronger than expected, the study of regime consequences would become easier. Research on the conditions of regime formation may also receive new momentum as a result of applying the logic of factor analysis to the variables included in the IRD. What conditions conducive to regime formation usually go together and which of these "independent variables" correlate negatively? Does this make it possible to identify different, but functionally equivalent, tracks or pathways leading to regime formation? Analyses of interrelationships between variables within individual sections of the IRD can produce answers to questions of this sort.

The uses of the IRD described in the preceding paragraphs cover a small selection of the overall possibilities. There is ample room for many analysts to make use of the IRD to probe a wide range of questions. We have no desire to limit the uses of the IRD to our own research programs. More importantly, the imagination and the energy of the whole community of regime analysts will help to exploit the potential of this database fully. We heartily invite all those who may be interested to make full use of this analytic tool. This is why we have included a CD containing both the IRD Data Protocol and the data themselves with each copy of this book.

### 6.3.3 Methodological Options

In this book, we have focused almost entirely on an examination of relationships occurring among variables included in the IRD and made use of descriptive statistics to explore the nature of these relationships. Thus, we have analyzed the roles institutional arrangements play as determinants of compliant behavior, the growth of knowledge about key problems, and problem solving. In so doing, we have worked with universes of cases that typically encompass several hundred individual records, and we have computed cross-tabulations as a means of determining the strength of the relationships between relevant variables. We are convinced that this is one appropriate way to make use of data included in the IRD. Certainly, it represents a major step forward from previous work on the effectiveness of international regimes.

Nonetheless, the IRD lends itself to studies that employ different methods of analysis. We have designed the database in such a way as to make it easy to import data on non-IRD variables and to link them to variables that are included in the database. If the analyst believes that democracies are more likely than other political systems to comply with the rules of international regimes, for instance, it is easy enough to import data pertaining to the political systems of member states and then to analyze the relationship between these data and the data on compliance included in the IRD. Similar comments are in order regarding the hypothesis that regimes whose members are advanced industrial societies (e.g., OECD members) are more likely to be successful in solving problems than regimes whose members are more diverse in economic terms. These comments are not meant to advocate any particular line of analysis. Rather, our purpose is to point to a use of the IRD that is different from the work described in chapters 3 to 5.

Beyond this, we note the opportunities to bring more powerful statistical tools to bear in analyzing the data included in the IRD. It would be feasible, for example, to treat problem solving as the dependent variable and then to analyze the relationships between scores on this variable and scores on a number of variables relating to regime attributes through the application of some form of multiple regression. Naturally, work of this sort would require efforts to solve specific methodological problems, such as the occurrence of multicolinearity among the independent variables included in the analysis. But there is nothing unusual about the data included in the IRD in these terms.

Conversely, the IRD offers a variety of opportunities for those desiring to make use of qualitative methods. Not only is it possible to disaggregate the data in order to examine individual regimes or even regime elements; the database also includes numerous text fields in which coders have added qualitative observations or narratives to accompany the answers they have provided in response to quantitative questions. Thus, the IRD questionnaire asks the coder to "describe the essential features of the regime's decisionmaking bodies as well as their interactions with each other." Similarly, it encourages coders to "elaborate on [the] compliance mechanisms provided for in the regime's constitutive provisions in a paragraph." Of course, the information supplied in response to queries of this kind is both qualitative and often couched in terms that are unique to the thinking of individual coders. Even so, the database includes a large amount of data of this sort that is readily accessible to those looking for new sources of qualitative information regarding the nature of regimes and how they work in endeavoring to solve international problems.

## 6.4 Future Directions

We have chosen to explore in this book a relatively narrow range of issues—the sources of compliance and the impacts of a few regime attributes on goal attainment and problem solving—in order to showcase the usefulness of the IRD in probing core questions relating to the roles that institutions play in international society. As a strategy for demonstrating how the database can be used to move the field of regime analysis to a new level of sophistication, we believe this choice has proven fruitful. Much remains to be done in the effort to deepen our understanding of the nature of regimes and the roles they play in international society. But this study represents a major step forward from the qualitative comparisons across relatively small numbers of discrete cases that have dominated regime analysis during most of its history.

One consequence of this strategy, however, is that we have barely scratched the surface in terms of bringing the data included in the IRD to bear on well-known and important questions about the nature of regimes and the roles they play in international society. Thus, we have said nothing about processes of regime formation, bypassed numerous issues regarding the significance of regime attributes, and examined regime consequences in a highly selective fashion. To the extent that we have succeeded in demonstrating the power of this information utility in the empirical chapters of this book, we hope this will stimulate others to think about additional uses of the database. Certainly, the IRD contains a large quantity of data that is readily accessible to anyone who has a serious interest in conducting research in this field.

Another result is more substantive in character. In our judgment, the work reported in this book leads to two propositions that will affect future work on the roles that regimes play in international society. First is the proposition that general debates about the relative importance of power versus interests versus ideas are not particularly helpful. We are convinced that making progress will require a focus on more specific causal mechanisms that often encompass elements of all three of the overarching drivers and that can explain outcomes in individual cases better than general appeals to the role of power, interests, or ideas as master variables.

As this observation suggests, a second proposition concerns the importance of contextualization. It is undoubtedly important to explore in a general way the effects of various regime attributes, such as decision rules, compliance mechanisms, revenue sources, administrative capacity, and so forth. But there is a critical need to focus as well on scope conditions rather than on all-encompassing generalizations. To take a concrete example, a logical next step regarding our findings relating to consensus as a decision rule would be to ask whether there are identifiable conditions under which consensus works well and others under which we should expect problems to arise from relying on the achievement of consensus.

Of course, scientific research is dynamic; new questions come into focus even as we struggle to develop convincing answers to existing questions. The study of international regimes is no exception in these terms. The effect of this is to limit the applicability of the IRD; the database does not contain data on topics that have come into focus since we designed the data protocol in the mid-1990s. Even so, this observation should not be allowed to obscure the fact that the IRD does contain large quantities of data dealing with a range of key ideas relating both to regime formation and to regime consequences.

In conjunction with the publication of this book, we are taking steps to make the IRD freely available to those interested in international institutions in general and environmental regimes more specifically. The files that make up the IRD are included in the CD accompanying this book. The database consists of a set of read-only files. This does not mean that we are opposed to adding new regimes and new data to the IRD. But such alterations need to be made in a manner that ensures compatibility with the existing data and that does not compromise the IRD as an information utility. Of course, we will be happy to engage in a dialogue with anyone who has ideas about ways to expand or restructure the database. In this way, we hope that our combined efforts can move the community toward a more sophisticated understanding of the roles that institutions play in international society.

# Appendix A
# Excerpts from the IRD Data Protocol

## PRECODING AGREEMENT

The International Regimes Database (IRD) is a computerized information system containing a wide variety of information on a continuously growing set of international regimes. The database is a research tool intended, in the first instance, for use by social scientists seeking to expand knowledge regarding the formation, effectiveness, and dynamics of regimes. The value of this knowledge to practitioners responsible for establishing and operating specific regimes should also be substantial.

### A.1 Database Architecture

The IRD data protocol contains four principal sections. Each section covers a grouping of variables that address a major area of interest to students of international regimes. The adoption of this structure is intended to facilitate the efforts of both coders and users. But users of the database can draw on variables located in different sections of the database in framing hypotheses to be explored on the basis of information contained in the IRD.

The first section of the data protocol deals with regime formation and places primary emphasis on understanding the forces at work (e.g., power, interests, ideas) in efforts to form new regimes. This emphasis on origins is understandable not only in light of the remarkable growth in the number and variety of regimes operative at the international level during the recent past but also in light of the conflicts and difficulties involved with long negotiation processes and deadlocks or even failures of regime building efforts. Regime formation, a subject that encompasses the reformation of existing institutional arrangements as well as the creation of new institutions where none has previously existed, has emerged as one of the central concerns of the "new institutionalism" in international relations.

The second section focuses on regime attributes and asks for data on principles, norms, rules, decision-making procedures, and programs of regimes. This section will yield an inventory of each regime's principal regulative, procedural, and programmatic elements. Regime analysts may use data on regime attributes to assess both the performance and the evolution of regimes. Policy-makers will be able to use quantitative data on regime attributes, such as compliance mechanisms, programmatic activities, decision rules, dispute settlement procedures, or organizational arrangements, to design effective international institutions. Regime analysts will also be able to comprehend changes in the attributes of regimes over time and to understand the evolution of attributes within individual regime components.

The third section on regime consequences contains variables dealing with domestic and international effects of regimes. Coders will be asked to provide data about the "real world" effects of regimes at the international and the national level. This section starts with the outputs of a regime, including activities involved in operationalizing the regime's provisions and decision-making procedures and the first official steps required to translate a regime's provisions from paper to practice. It then moves to data about outcomes at both the international and the domestic level including such variables as: what aims states formulated for problem-solving and goal-attainment in connection with the regime, compliance by member states, activities of major agencies, and actions of those affected by the regime's rules. The section also focuses on the impacts of a regime including both issue specific and broader results produced by the regime's operation. Such results encompass the regime's contribution to solving the problem(s) that motivated the parties to create it, the regime's contribution to learning about the nature of the problem as well as impacts on the distribution of values arising from the regime's activities.

The fourth section deals with matters of regime dynamics that are not captured in the first and third sections. Because the sections on regime formation and effectiveness allow coders to answer the same questions for the periods before and after fundamental changes (we call them watershed changes) in regimes, data pertaining to some aspects of institutional dynamics are included in these sections. As a result, section four is shorter than the other sections. But it does include data on a number of matters (e.g., shifts in the behavior of regime members toward the regime, resilience of the regime when confronted with new challenges) that are required to answer some important questions about the operations of international regimes.

## A.2 Precoding Questions

To afford maximum comparability among cases included in the database, we are asking you to complete a precoding agreement in consultation with members of the database management team. Among other things, this procedure will permit us to customize the computerized database protocol to suit the specific features of your case.

The case or unit of analysis for the International Regimes Database is the individual regime. A regime is a social institution consisting of agreed upon principles, norms, rules, decision-making procedures, and programs that govern the interactions of actors in a specific issue area. The scope of an issue area is subjectively defined, mainly on the basis of the perceptions of decision-makers and the organization of governmental structures within participating states. Regimes are social practices that may encompass both a number of explicit agreements and a variety of informal understandings that produce varying degrees of convergence of expectations regarding the behavior of regime members. As such, regimes have temporal boundaries in the sense of starting points and, in some cases, end points.

To reach agreement on the analytic structure of an individual case, it is necessary to identify: (1) the problem(s) to which the regime constitutes a response, (2) the external boundaries of the regime or, in other words, what is part of the regime and what is external to it or what constitutes the environment in which it operates, (3) the internal boundaries of the regime, including discrete components of the regime itself and the temporal structure of the regime, and (4) the most important actors in the regime's issue area.

Although it is not strictly a questionnaire, this precoding agreement is framed as a series of questions. You should treat them as the basis for a dialogue between yourself and members of the database team. Please draft initial responses to the questions and forward them to the database manager by email or fax. If any question is unclear to you, please contact us right away. Once the precoding agreement is finalized, the database manager will customize the electronic version of the database protocol to reflect the terms of this agreement for your case. Note that for purposes of this precoding agreement, the case needs to be outlined, not described in great detail. To keep the task of coding a case manageable, it is desirable to limit the number of regime components as much as possible and to aggregate the important actors as far as possible.

### A.2.1 The Problem

International regimes are generally problem driven in the sense that they come into existence to solve or manage problems (e.g., jurisdictional conflicts, transboundary fluxes of pollutants, health hazards) that individual countries are unable to cope with on their own. Yet there is often considerable scope for interpretation in framing these problems; participants do not always agree with each other regarding the nature of the problem, and specific regimes can deal with more than one subproblem at the same time.

**A.2.1.1 Problem Definition** What problem stimulated participating actors to form the regime? If possible, describe the problem in terms of activities causing the problem or in terms of costs—of any kind—attributable to the problem. If it is hard to answer this question unambiguously, describe two or more problems underlying the regime and specify whether these are subproblems representing distinct elements of the issue area or alternative ways of framing the problem on the part of different actors. Identify only those subproblems or alternative formulations that are consequential for regime formation, design, or effectiveness.

Examples: The regime dealing with trade in chemicals and pesticides illustrates the case of subproblems. It covers two subproblems, one pertaining to public health and the other dealing with the harmonization of international trade. The Baltic Sea regime, by contrast, illustrates the case of differing perceptions of the problem, since some participants framed the problem as one of marine pollution while others saw it as a matter of finding a vehicle for demonstrating the possibility of East-West cooperation.

Notes. Question 2.1.1 pertains to the period of regime formation. You will be able to deal with subsequent changes in the nature or formulation of the problem at a later stage in the coding of your case. Also, do not finalize your answer to this question before you have considered question 3.1.1

### A.2.2 External Regime Boundaries

Although regime formation is not an instantaneous process, it is useful for purposes of analysis to specify a temporal starting point for each case included in the database. Many regimes operate on an ongoing basis. But international regimes can go out of existence. When this happens, we want to specify a termination point for the regime. In addition to these temporal boundaries, it is important to specify the substantive boundaries of regimes in the sense of differentiating between elements of the regime itself and features of the surrounding environment. Normally this involves identifying the regime's constitutive elements, whether these are formal agreements or informal understandings.

**A.2.2.1 Regime Formation** When did regime formation occur in your case? Is there more than one plausible starting date for this case? If so, please provide a short account of the relative merits of each possible starting point and indicate your preference. We normally date the establishment of a regime from the signing of a document (whether legally binding or not) in which the parties agree to the constitutive provisions of an institutional arrangement.

**A.2.2.2 Regime Termination** Did the regime cease to exist? A regime ceases to exist if (i) there is a temporal discontinuity in the operation of the regime and (ii) there is no identifiable successor (iia) or (iib) the successor involves a fundamental change in terms of leading actors or the definition of the problem leading to new governing principles/key norms that are at odds with the old ones. Temporal discontinuity occurs when (ia) a complete cessation of regime activities occurs or (ib) actors no longer feel obliged to justify actions that contradict regime obligations or (ic) one or more essential parties abrogate the agreement. Specify the date and describe the nature of the termination process, taking the earliest indicator for regime discontinuity.

Notes. This is a high threshold. To illustrate, the North Pacific fur seal regime came into existence long before World War II, but did not operate during the war years. Because it was revived after the war without significant changes in leading actors or in guiding principles/norms, we analyze it as one regime. Similarly, despite the adoption of governing principles at odds with those articulated in the 1946 convention, we treat the whaling regime as one case because there was no discontinuity in the operation of the regime. If we have asked you to code a regime that strikes you as two distinct cases on the basis of these criteria, please contact us immediately. If your case involves significant changes that fall short of our criteria of regime cessation, you will have an opportunity later on to identify watershed changes within the regime.

**A.2.2.3 Constitutive Elements** Identify all agreements that you regard as constitutive elements of the regime by name and date (i.e., when they came into existence and dates of major revisions), whether they are (i) legally binding agreements, or (ii) explicit agreements that are not legally binding (i.e., soft law), or (iii) de facto practices that are not formulated in any written agreement (i.e., tacit rules). Where relevant, state whether these agreements are in force.

### A.2.3 Internal Regime Boundaries

We have found that it is also helpful to make distinctions among analytically differentiable components and temporal watersheds within the same regime. These distinctions are not as fundamental as those described in the preceding section; they pertain to differences within regimes rather than between regimes. Yet it is often necessary to make use of such distinctions to formulate intelligible answers to questions included in the database protocol.

Many regimes encompass two or more distinct institutional arrangements—we call them components—that together make up the regime as a whole. Many questions included in the database protocol apply to individual components rather than to the overall regime. In some but not all cases, it is possible to identify a single component (e.g., a framework convention) as the regime's core with other components (e.g., substantive protocols) occupying positions ancillary to the core.

Note. Because the existence of components adds substantially to the work of coding a case, you will want to be cautious in identifying components in your case. To guide your thinking, we have developed the following criteria:

Regimes have separate components if there are (i) distinct institutional forms (e.g., separate treaties, protocols, annexes) and (ii) one or more of the following conditions obtains:

(iia) the distinct forms deal with different subproblems (see above);

(iib) they cover differentiable sources of the problem(s) (e.g., the European acid rain regime includes separate protocols regulating emissions of $SO_X$, $NO_X$, and VOCs in addition to its framework convention);

(iic) these forms are aimed at different regulatory targets (e.g., the global trade regime encompasses different rules for trade in manufactured goods, agricultural goods, and services),

(iid) they aim at different clusters of actors (e.g., the nuclear nonproliferation regime includes different norms and rules aimed at nuclear-weapon states and non-nuclear-weapon states).

(iie) these forms deal with major regime functions (e.g., compliance, funding).

**A.2.3.1 Regime Components** Does the regime have two or more components? Which of the agreements identified under 2.3 are applicable to each component? List by component.

**A.2.3.2 Component History** When did each component come into existence and become a part of the regime? Provide specific dates.

**A.2.3.3 Regime Core** Does one of these components constitute the core of the regime? If so, identify the core.

**A.2.3.4 Watershed Change** A watershed, in contrast to a transition from one regime to another, marks a major change within a regime and divides the regime into distinct time periods. A watershed occurs if there is (i) a temporal discontinuity in the operation of the regime, or (ii) a significant restructuring of principles/key norms (the new principles need not necessarily be at odds with the old ones), or (iii) a significant change in the group of leading actors, or (iv) a significant expansion in functional scope (e.g., a radical deepening of regulative rules). In cases involving watersheds, you will be asked separate questions covering the periods before and after the watershed change.

Did the regime experience one or more watersheds? If so, specify the date(s) and the major causes and circumstances. Indicate which of the criteria listed above are relevant to your judgment.

If a watershed change occurred, was it accompanied by a change in the nature of the problem(s) or subproblems or in the framing of the problem(s)? Explain how the problem(s) after the watershed differ from the problem(s) mentioned under 2.1.

## A.3 Important Actors

Many parts of the protocol take an actor-oriented perspective. For instance, we often ask about activities instead of structures in seeking to identify the causes of the problem(s). The problem addressed by the GATT regime, for example, is treated at the outset as a problem arising from the protectionist policies of states rather than as a collective-action problem. Similarly, we start by asking about activities leading to overfishing rather than by characterizing the problem as a tragedy of the commons. Later on, we ask additional questions dealing with structural conditions.

In cases where there are only a few actors, it is possible to supply data on all those participating in the formation and operation of a regime. A growing number of cases, however, involve too many actors to cover exhaustively. In such cases, it is important to identify the most important actors for purposes of answering actor-specific questions.

In some but not all regimes, important actors include not only states and coalitions of states but also non-state actors and even individuals (when their influence reaches beyond their role as agents of corporate actors). Note that an actor can be counted as important whether or not it is a member of the regime (e.g., its actions may be a major cause of the problem even if it does not become a regime member). Disaggregation with regard to actors adds to the complexity of the database protocol and should be resorted to sparingly. If stable coalitions are important in your case, identify them. But also identify one or two leading states for each coalition.

Add a sentence or two indicating why each of these actors is important. Factors to consider in answering this question include: (i) major role in causing the problem(s), (ii) major role (potential or actual) in solving the problem(s), (iii) major role in creating the regime, and (iv) major role in developing strategies to implement the provisions of the regime.

### A.3.1 Important Actors

Which states, coalitions, non-state actors, or individuals do you regard as the most important for purposes of answering questions relating to your case that are actor-specific? Provide separate answers for each of the following subquestions.

**A.3.1.1** If applicable, identify up to five or six important states.

**A.3.1.2** If applicable, identify important state coalitions. Identify one or two leading states for each coalition.

**A.3.1.3** If applicable, identify up to five or six important non-state actors.

**A.3.1.4** If applicable, identify up to five or six important individuals.

## IRD DATA PROTOCOL—TABLE OF CONTENTS

### PART I—Regime Formation

#### 101 The Problem

101A How many nations were regarded as being important because of their role in causing the problem? How many were regarded as being especially important compared to others because of their role in causing the problem?

101B How many nations were regarded as being important because of the extent to which they were affected by the problem? How was the damage associated with the problem shared among these negatively affected?

101C Which of the important nations identified in the precoding agreement and which other nations were considered to be important because of their roles in causing the problem and/or the negative effects of the problem?

101D Compare the set of nations especially important because of their roles in causing the problem with the set experiencing the bulk of the suffering.

101E What current or future negative effects did or could the problem(s) have on these nations? Specify whether negative effects result in a loss of current welfare or in a failure to achieve gains. Distinguish whether these losses are currently occurring or are expected to occur in the future. Rank up to five of the most important effects.

101F Regarding interests involved in the issue area: Did the negotiations regarding the regime deal with conflicts about the distribution of the costs?

101G Regarding interests involved in the issue area: Was there an incentive to disobey the rules even after the regime was put in place?

101H Regarding interests involved in the issue area: How complex was the issue area?

101I Regarding interests involved in the issue area: How compatible/incompatible were the interests of the parties?

101J Regarding interests involved in the issue area: What was the degree of symmetry/asymmetry in terms of present versus future impacts of the problem?

101K Describe the nature of the proposed solutions to the problem(s). If states had different views about alternative solutions, please explain.

101L Did the proposed solutions appear to be difficult or easy to implement?

101M What were the most important difficulties in implementing the proposed solutions?

101N Does the problem involve supplying a collective good, regulating the use of a common pool resource, managing a shared natural resource, or controlling transboundary externalities?

101O Does the problem involve a conflict over values, a conflict of interest about relatively assessed goods, a conflict over means, or a conflict of interest about absolutely assessed goods?

101P Is the problem addressed by the regime relatively self-contained or closely linked to surrounding issues?

**102 The Political Setting**

102A What broader currents of tension and hostility or friendship prevailed among the parties participating in regime formation?

102B What power resources are relevant to determining the distribution of issue-specific power resources among the participants in regime formation? List the three most powerful states having these (or most of these) power resources.

102C Were the nations involved in regime formation roughly symmetrical in terms of issue-specific power or did the process involve sharp differences in power resources?

**103 The Economic Setting**

103A What is the relative importance of the economic sectors that dominated the issue area for the single nations involved? Indicate importance for sectors directly or indirectly responsible for and/or causing the problem(s). Indicate relative importance of economic sectors for each important nation identified in the precoding negotiations.

103B How were costs and benefits of efforts to solve the problem distributed among nations?

**104 The Cognitive Setting**

104A Was the nature of the problem well understood?

104B How complete was the information about the options available for dealing with the problem?

104C Did the parties disagree about the completeness or accuracy of information about the options?

104D How well established was knowledge of the probable consequences of different options for solving the problem?

104E Did the parties disagree about the probable consequences of selecting different options?

104F What expert groups were present and active during the process of regime formation?

**105 States during Regime Formation**

105A For each of the important states identified in the precoding agreement, indicate whether that state was a pusher, a laggard, or neutral. Include other states if especially noteworthy.

105B For each of the important states identified in the precoding agreement and above, indicate whether that state's efforts were primarily structural, ideational, or entrepreneurial.

105C Were the efforts of individual pushers reinforcing or conflicting? Were the efforts of individual laggards reinforcing or conflicting?

**106 Non-state Actors during Regime Formation**

106A What types of non-state actors were actively involved?

106B For each of the important non-state actors identified in the precoding agreement, indicate whether that non-state actor was a pusher, a laggard, or neutral. Include other non-state actors if especially noteworthy.

106C For each of the non-state actors identified in the precoding agreement and above, indicate whether its efforts were primarily structural, ideational, or entrepreneurial.

106D Were the efforts of individual non-state pushers reinforcing or conflicting? Were the efforts of individual non-state laggards reinforcing or conflicting?

**107 Individuals during Regime Formation**

107A For each of the important individuals identified in the precoding agreement, indicate whether that individual was a pusher, a laggard, or neutral. Include other individuals if especially noteworthy.

107B For each of the individuals identified in the precoding agreement and above, indicate whether her/his efforts where primarily structural, ideational, or entrepreneurial.

107C Were the efforts of individuals who were important pushers reinforcing or conflicting? Were the efforts of individuals who were important laggards reinforcing or conflicting?

**108 Agenda Formation**

108A When did the problem to be addressed by the regime initially make its way onto the international political agenda? Did the problem stay on the agenda continuously until the negotiations started? When was the last time the problem appeared on the agenda before the negotiations started?

108B Which of the following factors were present during agenda formation? Which of these factors were most influential for agenda formation?

108C Provide a textual explanation of what caused the shift from agenda formation to explicit negotiations.

108D Was the inclusion of issues on the agenda determined by a single state or a small group of potential regime members? Was the inclusion of issues on the agenda determined by potential regime members negotiating among themselves more or less as equals and setting the terms of the agenda deliberately? Was the inclusion of issues on the agenda determined by factors largely outside deliberate efforts of potential regime members?

**109 Negotiation Stage**

109A Were there earlier negotiations that failed to create a regime in the issue area? If yes, when did the latest effort that failed start and end?

109B When did the explicit negotiations that led to the regime start? When did they end?

109C Under whose auspices did the negotiations take place?

109D How many actors participated in the negotiations as official participants and observers?

109E Were there states not participating in the negotiation process that other actors believed should have participated?

109F Indicate the negotiation strategies employed by each of the important states and coalitions identified in the precoding agreement. Write a paragraph describing the essential nature of the strategies used during the negotiation phase employing the keywords checked below.

109G Did potential signatories participate in the negotiations but fail to sign the agreement reached?

109H What roles did non-state actors play in the negotiations?

109I In a short paragraph, evaluate how influential these non-state actors were during the negotiations.

109J Were the negotiations dominated by a single state or a small group of states, by an interstate process among relative equals, or by transnational forces?

**110 Operationalization Stage**

110A What steps were required for the relevant agreements to become operational? Indicate these steps for each agreement identified in the precoding agreement.

110B How long did it take for the relevant agreements of the regime to become operational? Provide information for each relevant agreement identified in the precoding agreement.

110C Were there serious disagreements among potential regime members concerning whether to make the agreement(s) operational? If so, were these disagreements resolved through actions of a single state or a small group of states, by an interstate process among relative equals, or by transnational forces?

110D Did the parties agree to act as if the agreement were already in force before the regime became operational?

110E How severe were the domestic political battles over the terms of the agreement within member states during the operationalization stage?

### 111 Narrative

111A Write a short description of the central aspects of regime formation. It is acceptable to repeat information coded above if you find it helpful. It is also appropriate to include information not coded above. List the most important literature used in coding the regime formation part of the data protocol.

## PART II—Regime Attributes

### 201 Stated and Unstated Goals

201A Does the regime have stated goals? If so, what are the most important stated goals of the regime? Identify up to five.

201B Does the regime have unstated goals? If so, what are the most important unstated goals of the regime? Identify up to five.

### 202 Larger Vision or Theory

202A Does the regime reflect some larger vision or theory regarding the organization and operation of international society?

### 203 Overall Characteristics

203A Is the regime an explicit or a tacit regime?

203B To what extent is the regime an internal or an external regime?

203C Do the principles and norms of the regime apply only to members as such, or are members responsible for ensuring that these provisions are complied with by other actors operating within their jurisdiction?

203D Can the regime's principles be characterized in terms of the three fundamental types of property rights/allocation rules?

## 204 Principles and Norms

204A Identify the most important regime-specific principles and norms. Mention whether the constitutive agreements make explicit reference to each principle or norm, or whether the commitment is implicit.

## 205 Regime Rules

205A What are the substantive rules or prescriptions of the regime? Indicate whether the constitutive agreements make explicit reference to the rules.

205B Is it possible to categorize the regime's substantive rules as requirements, prohibitions, or permissions?

205C Are the regime's substantive rules legally binding on the members, or do they have the character of soft law (e.g., ministerial declarations, codes of conduct)?

205D Are the regime's substantive rules generally precise and easy to interpret in the sense that they call for well-defined actions, or are they ambiguous and indeterminate?

205E Does the regime have substantive rules that differentiate among its members in terms of requirements, prohibitions, or permissions?

205F Is the regime narrow or broad as measured by the functional scope of its rules?

205G Is the regime shallow or deep as measured by the density and specificity of its rules?

## 206 Programs

206A Does the regime explicitly call for the conduct of programmatic activities?

206B If the regime explicitly calls for the conduct of programmatic activities, to what extent are those who participate in relevant advisory bodies/programmatic activities recognized as experts in the field?

206C If the regime explicitly calls for the conduct of programmatic activities, who controls participation and the agenda in the bodies carrying out programmatic activities?

206D Does the regime have de facto programs that differ materially from explicitly mandated programs?

**207 Membership**

207A How many formal members do the relevant agreements have?

207B Are there informal members (states or non-state actors that have not formally signed the regime's constitutive agreements or otherwise acquired membership) but that have rights and obligations under the terms of the regime and have a voice in its decisionmaking processes? If yes, identify these members.

207C Are there state or non-state actors that are important in terms of the problem to be solved but that are not members of the regime?

207D Are there explicit provisions that target activities (e.g., oil tanker operators in MARPOL) of non-state actors or that cover the participation of non-state actors (e.g., provisions governing observer status for non-state actors in the Convention on International Trade in Endangered Species of Wild Flora and Fauna) in the regime? If so, elaborate on the nature of the rules in a paragraph.

**208 Membership Criteria and Rules**

208A What criteria govern eligibility for membership?

208B How many states meet these criteria?

208C Do the rules pertaining to the admission of non-founding members differ from the admission rules applied to founding or original members? If yes, please describe.

208D Is there a single category of membership or are there provisions establishing more than one category of membership?

208E Do the regime's provisions allow for role differentiation among the members?

**209 Secretariat**

209A Did the members of the regime establish a secretariat for the regime as a whole or any of its elements?

209B How independent is the secretariat from the regime's members?

## 210 Decision Making

210A What types of formal decisions are made by the regime?

210B What decision rules does the regime provide for and use in arriving at decisions?

210C What decisionmaking bodies are provided for in the regime?

210D How do these bodies participate in decisionmaking? Is participation by these bodies carried out in accordance with the constitutive provisions of the regime?

210E Are there other bodies within the regime that play a role in decisionmaking but are not explicitly provided for in the regime's constitutive provisions? How do these bodies participate in decisionmaking?

210F Does the regime formally or informally delegate decisionmaking power to any external bodies?

210G How do the external bodies to which the regime informally or formally delegated decisionmaking power participate in decisionmaking?

210H Are there substantive or procedural restrictions affecting the issues that get on the decisionmaking agenda of bodies explicitly provided for by the regime?

210I Are there substantive or procedural restrictions affecting the issues that get on the decisionmaking agenda of bodies not explicitly provided for by the regime?

210J Are there substantive or procedural restrictions affecting the issues that get on the decisionmaking agenda of external bodies to which the regime formally or informally delegates decisionmaking power?

210K In a short paragraph describe the essential features of the regime's decisionmaking bodies as well as their interactions with each other.

## 211 Funding Mechanisms

211A How are the regime's activities and administrative operations (including the work of the secretariat) funded?

211B How are the programmatic activities of the regime funded?

211C What kinds of financial mechanisms have been created under the regime? Does the regime make use of annual or multiyear budgets?

### 212 Compliance Mechanisms

212A Are there reporting procedures requiring the submission of information by individual members pertaining to regime implementation?

212B Are there procedures for reviewing implementation formally or not formally established in the regime's constitutive provisions?

212C What formal compliance mechanisms are provided for in the regime's constitutive provisions to achieve compliance?

212D What compliance mechanisms not provided for in the regime's constitutive provisions are used to achieve compliance?

212E Do these procedures generally reflect an enforcement approach or a management approach to compliance?

### 213 Regime's Interactions with Other Institutions

213A Identify other regimes with which this regime interacts. Indicate whether the interaction is mutually reinforcing or conflicting.

### 214 Narrative

214A Write a short description of the central aspects of the regime's attributes. It is acceptable to repeat information coded above if you find it helpful. It is also appropriate to include information not coded above. List the most important literature used in coding the regime formation part of the data protocol.

## PART III—Regime Consequences

### 301 Outputs at the International Level

301A Are the regime-wide bodies that are called for by the regime's constitutive agreements in operation? If so, do these regime-wide organizational arrangements produce authoritative decisions on a regular or as-needed basis?

301B What kinds of authoritative decisions about principles, norms, rules and programmatic activities are made by the regime?

### 302 Member-level Outputs

302A Have the important members taken steps to translate the international commitments of the agreements into domestic obligations?

302B In a paragraph, elaborate on the political dynamics involved in these outputs.

### 303 Actor-level Outcomes

303A Does the behavior of important actors generally conform with the provisions of the regime? Did the regime exert a causal influence on these developments?

303B What types of events and actions were significant elements of the processes through which the regime affected outcomes? Which of these processes played significant roles in the regime's causal impact?

303C Did the regime have behavioral effects that were not explicitly called for in its constitutive provisions? If so, please describe these effects and the mechanisms that caused them.

303D Has the operation of the regime directly or indirectly affected the behavior of various social groups of important actors (e.g., car drivers, power plant operators, builders of pollution abatement facilities, private firms, ethnic groups, arms producers)?

303E Specify the important nations in which the behavior of these groups was especially affected. Indicate the extent to which the behavior of these groups was affected by the operation of the regime.

303F Taken together, did the behavioral changes lead to the fulfillment of the stated and/or unstated goals of the regime? What causal impact did the regime have in producing these changes?

### 304 Impacts of the Regime in the Targeted Issue Area

304A How did the state of the world change during this period with respect to the problems addressed by the regime? Did the regime exert a causal influence on these developments?

304B Did the understanding of the nature of the problem change within important nations? How much of this change is attributable to the operation of the regime?

304C Did the information about the options available for tackling the problem change within important states? How much of this change is attributable to the operation of the regime?

304D Did the operation of the regime lead to increases in the capacities of member states to participate effectively in social practices at the international level? If so, elaborate in a paragraph.

304E How are the benefits arising from the operation of the regime distributed among members? In a paragraph, describe the nature of the distribution and identify a few illustrative states that received different benefits.

304F How are the costs associated with the operation of the regime distributed among members? In a paragraph, describe the nature of the distribution and identify a few illustrative states that incurred different costs.

304G Describe the relationship between the distribution of costs and benefits.

**305 Broader Consequences of the Regime**

305A Did the regime influence relations between members and nonmembers? If so, please describe.

305B Did the operation of the regime produce demonstration effects leading to the creation of new international regimes or influencing the evolution of existing international regimes in other issue areas? If so, please describe.

305C Has the regime changed the contents of the international political agenda or the priority of issues included on this agenda?

305D Has the regime affected overall relations among the participating states?

305E Has the regime affected the character of international society?

**306 Narrative**

306A Write a short description of the central aspects of the regime consequences. It is acceptable to repeat information coded above if you find it helpful. It is also appropriate to include information not coded above. List the most important literature used in coding the regime consequences part of the data protocol.

## PART IV—Regime Dynamics

### 401 Changes in Connection with Watersheds

401A If a watershed occurred, what kinds of changes did it involve? Indicate what caused the change to occur.

### 402 Dynamic Patterns at the Overall Level

402A Did any challenges to the survival of the regime occur?

402B Describe the dynamic patterns at the overall level of the regime. What causes of these dynamic patterns at the overall level are identifiable?

402C Did the ability of the regime to cope with the problem(s) it addresses increase or decrease over time?

### 403 Narrative

403A Write a short description of the central aspects of the regime dynamics. It is acceptable to repeat information coded above if you find it helpful. It is also appropriate to include information not coded above. List the most important literature used in coding the regime dynamics part of the data protocol.

# Appendix B

# Using Formal Concept Analysis to Explore the International Regimes Database

Tim B. Kaiser

## B.1 Introduction

Formal Concept Analysis (FCA) is a mathematical theory that formalizes the philosophical understanding of a concept as a unit of human thought. Today, we can see Formal Concept Analysis embedded in the general philosophical framework of Contextual Logic and Conceptual Knowledge Processing (Stumme and Wille 2000), where mathematical techniques are developed in order to support humans in their autonomous thinking processes. Therefore, methods that keep a transparent connection to the underlying data are preferred. One such method is FCA, which makes it possible to transform given data into a diagrammatic representation that makes it easier to see hidden structures and dependencies. This is done in a transparent way by using statistical methods that break down the original data into numbers that do not allow for full reconstruction of the original data.

FCA has been applied successfully in various research areas, such as linguistic (Großkopf and Harras 2000), medical (Scheich et al. 1993), and political analysis (Kohler-Koch and Vogt 2000), for information retrieval (Rock and Wille 2000), as well as for On-Line Analytical Processing (Stumme 2000) and Knowledge Discovery in Databases (Hereth et al. 2000; Hereth et al. 2003).

The theory allows the analyst to derive conceptual hierarchies from data. These hierarchies can be visualized by line diagrams that stimulate human reasoning and activate geometric thinking. The need for computational support for applying these techniques gave rise in the early 1990s to the idea of a TOSCANA system consisting of different software components. Through the interface part of the system provided by the TOSCANA software, users can navigate graphically through conceptual hierarchies. They can explore and analyze a conceptual landscape derived from a connected database and a conceptual file. The conceptual file contains a professional conceptualization of the domain of the database, usually elaborated by a domain expert. The latest implementation of TOSCANA—used for the IRD-TOSCANA system—is written in the platform-independent programming language JAVA, allowing the system to run on Windows and Unix machines as well as on a Macintosh, assuming that the JAVA runtime is installed.

In this appendix, I illustrate the use of FCA as a tool for analyzing data in the IRD. The first section provides a short survey of the basic techniques in Formal Concept Analysis (sacrificing mathematical exactness for the sake of the needs of a broader audience). The second section describes the general structure of a TOSCANA system, using examples from the IRD to demonstrate the value and epistemological scope of this type of analysis. In the third section, I report on the development process of the IRD-TOSCANA system, emphasizing both benefits and problems. The final section describes special features of the TOSCANA system relevant to the IRD project and provides examples. Finally, I summarize the discussion and draw some conclusions of interest to users of the IRD.

## B.2 Formal Concept Analysis

Inherent in the mathematical theory of Formal Concept Analysis is the idea of a concept as a dyadic entity consisting of an *extent* and an *intent*, where the extent is the set of all objects falling under the scope of the concept and the intent is the set of all attributes characterizing these objects. For example, the concept "international regime" may have as its extent the set of all existing regimes and as its intent the set {"is constructed by humans", "at least two nations are involved", "is an institution"}. From this example we can see that it is not possible to describe the extent and the intent of a concept in general because our selection of attributes is arbitrary {why not list also "is not a flower"} and because our set of objects (all international regimes) cannot be captured easily.

As a result, we have to restrict our interest to a certain piece of reality. That is, we have to limit our focus to a defined set of objects and a defined set of attributes. Then, we have to decide which attribute applies to which object. From such a specified (formalized) part of reality, called *formal context*, we can deduce concepts in a mathematical and reproducible fashion. We can order these concepts in a hierarchical way, where a more specific concept is smaller (because fewer objects are in its extent) than a more general concept. For example, the concept "international environmental regime" is more specific (less general) than the concept "international regime." In this system of analysis, therefore, the concept "international environmental regime" is on a lower hierarchical level (*smaller*) than "international regime." The ordered set of concepts of a formal context is called *concept lattice*.

I am now able to provide an example constructed from the IRD to exemplify FCA. First, we consider a formal context represented by a data table (table B.1).

Here the set of objects consists of the components of the IATTC regime together with their coder (this doubling of the regime components by linking the coders is discussed in detail in section 3). The attribute set contains three programmatic activities in the area of management. A cross in a cell in the table indicates that the object in its row performs the activity given in the corresponding column head; for instance, the regime component *Conservation and Management of Tuna and Tuna-like Fishes 1949–1976* performs *information management* according to the coder James Joseph.

From this data table, we can derive the conceptual hierarchy shown in figure B.1.

In the following paragraphs, I discuss how to understand the diagram in figure B.1, which represents the concept lattice of the formal context in table B.1. Every node represents a *formal concept*. The lines visualize the *subconcept-superconcept* relation. The labels positioned slightly above a node list attributes and the labels slightly below a node list objects. The extent of a formal concept represented by a node consists of all objects that can be reached by downward line paths and dually its intent is the set of all attributes that can be reached by upward paths. The circle on the right-hand side, for instance, represents the concept consisting of the

**Table B.1**
Formal context "Programmatic Activities: Management"

| | Information management | Reviewing of adequacy of commitments | Financial and technology transfer |
|---|---|---|---|
| Conservation and Management of Tunas and Tunalike Fishes, 1949–1976, James Joseph | | | |
| Conservation and Management of Tunas and Tunalike Fishes, 1949–1976, Virginia Walsh | | | |
| Conservation and Management of Tunas and Tunalike Fishes, 1976–1998, James Joseph | X | | X |
| Conservation and Management of Tunas and Tunalike Fishes, 1976–1998, Virginia Walsh | X | | X |
| Conservation and Management of Dolphins, 1976–1998, James Joseph | | X | X |
| Conservation and Management of Dolphins, 1976–1998, Virginia Walsh | | X | X |

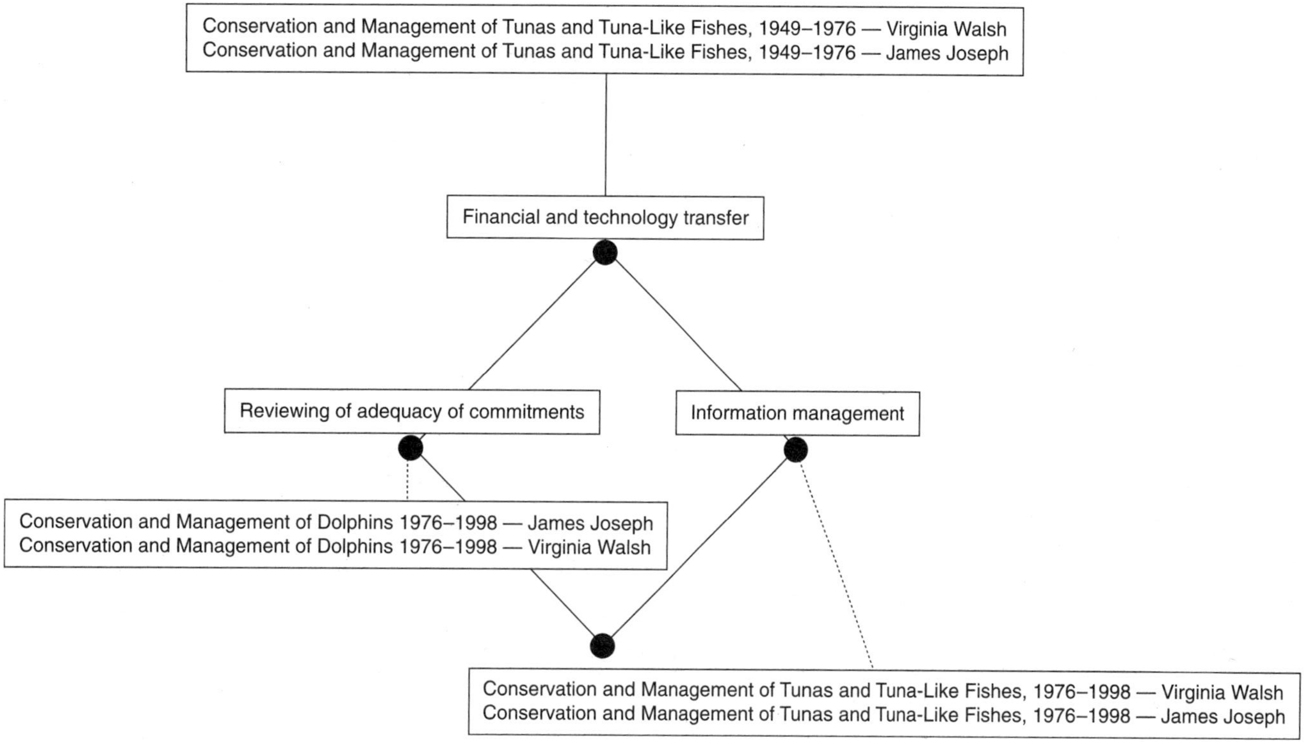


**Figure B.1**
Line diagram for a concept lattice of the formal context from table B.1

object set {Conservation and Management of Dolphins, 1976–1998, Virginia Walsh; Conservation and Management of Dolphins, 1976–1998, Joseph James} together with the attribute set {reviewing adequacy of commitment, financial and technology transfer}.

It is important to note that the concept lattice allows for reconstruction of the formal context on which it is based. That means that it completely represents the binary data. Additionally, we can easily detect linkages between attributes in the data table. For instance, an attribute implies another attribute if it can be reached by an upward path of line segments. In our example we see that every regime component that performs *Information management* also performs *Financial and technology transfer*. For a more detailed introduction to Formal Concept Analysis, see Ganter and Wille 1999 or visit the Formal Concept Analysis Homepage at http://www.upriss.org.uk/fca/fca.html, where pointers to many Formal Concept Analysis resources are given.

### B.2.1 Conceptual Scaling

Concept hierarchies as described above can only be derived from two-valued data. But in real life, we often are confronted with many-valued data like those included in the IRD. To make use of line-diagram representations of concept hierarchies hidden in many-valued data, we have to translate them into binary form. In formal concept analysis, this is done via *conceptual scaling* (Ganter and Wille 1989). A set of many-valued attributes is translated into a certain number of binary (yes-no) attributes by a specified rule. Such a rule for translating many-valued attributes can be represented by a formal context, called *conceptual scale*, where the set of objects contains the value combinations of the attributes to be translated. We can transform the scaled many-valued attributes into binary ones by exchanging the attributes from their conceptual scale for these attributes. Then, an object has an attribute if the previous value combination of the object has this attribute in the conceptual scale. We can illustrate this for a conceptual scale for the attribute *management programmatic activities*. We use the following abbreviations: Information Management = IM, Reviewing of adequacy of commitments = RAC, Financial technology and transfer = FTT.

**Table B.2**
Conceptual scale

| | IM | RAC | FTT |
|---|---|---|---|
| IM | X | | |
| RAC | | X | |
| FTT | | | X |
| IM and RAC | X | X | |
| RAC and FTT | | X | X |
| IM and FTT | X | | X |
| IM, RAC, and FTT | X | X | X |
| No activities | | | |

If we use the conceptual scale from table B.2 for translating the many-valued attribute *Programmatic Activities: Management* included in table B.3, we get as a result the formal context in table B.1.

If we have a conceptual scale available for every attribute of a many-valued context, we can translate the many-valued data into binary-valued data by applying the previously described process to every attribute. Since it is binary, we can build a concept lattice from the resulting *derived context*.

## B.3 How to Understand and Use the TOSCANA System for IRD Analysis

The idea of a TOSCANA system is a result of long experience in data analysis using concept lattices for deeper understanding and proper communication of data. The use of conceptual scaling combined with computational support yields a powerful analytic tool. The idea of the basic structure of such a system was first pointed out in Vogt, Wachter, and Wille 1991, Wille 1992, and Scheich et al. 1993.

The main components of the system consist of TOSCANA software, a conceptual file, and a database that together form a TOSCANA system (figure B.2).

The conceptual file contains the conceptual scales and corresponding line diagrams for attributes of a data table drawn from the database.

**Table B.3**
Many-valued context/data

| | Programmatic activities: management |
|---|---|
| Conservation and Management of Tunas and Tunalike Fishes, 1949–1976, James Joseph | No activities |
| Conservation and Management of Tunas and Tunalike Fishes, 1949–1976, Virginia Walsh | No activities |
| Conservation and Management of Tunas and Tunalike Fishes, 1976–1998, James Joseph | IM and FTT |
| Conservation and Management of Tunas and Tunalike Fishes, 1976–1998, Virginia Walsh | IM and FTT |
| Conservation and Management of Dolphins, 1976–1998, James Joseph | RAC and FTT |
| Conservation and Management of Dolphins, 1976–1998, Virginia Walsh | RAC and FTT |

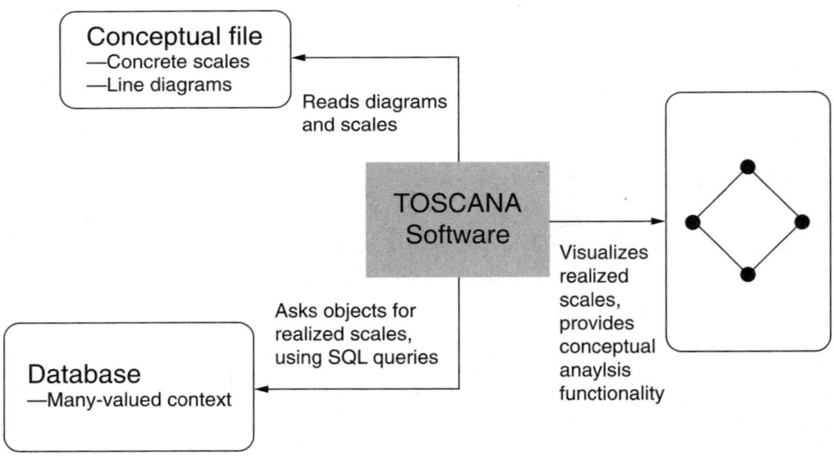


**Figure B.2**
Architecture of a TOSCANA system

The software, called TOSCANA, links the scales from the conceptual file to the database, thereby enabling the user of the system to explore the conceptual structure of the data by *displaying* line diagrams of concept lattices of pieces of the data. Additionally, the system provides two core functionalities as support for the analysis process: *nesting* and *zooming*. In the following, I describe the functionality of the system in detail.

### B.3.1 Diagram Display

In the IRD-TOSCANA system, we see—instead of a line diagram of a concept lattice as in figure B.1—diagrams of the type shown in figure B.3.

The concept lattice represented in figure B.1 is contained as a substructure in the line diagram in figure B.3. It is marked by the big circles. The line diagram in figure B.3 represents the concept lattice of the conceptual scale shown in table B.2 but with the former objects (which have been attribute values) replaced by our objects of interest from the IRD. Because the many-valued context shown in table B.3 does not include all possible values, we get *nonrealized concepts*. For instance, the concept represented by the small circle labeled *Information management* is nonrealized, since there is no object in the many-valued context with the value *IM* and also no smaller concepts combine to form a new one at this node (which is the case on the opposite site: the concept labeled *Financial and technology transfer* is represented by a big circle that shows that the corresponding concept is realized, since its object set consists of all four objects that can be reached by following downward line paths). We can make this explicit by changing the labeling option to *show all matches* (figure B.4).

Thus, the nonrealized concepts tell us what types of objects do not exist in the database but may exist from the viewpoint of the designer of the conceptual scale. This indicates a gap between the theory the developer of the scale had in mind and the empirical data. The explanation for this could be an incorrect theory or just an incomplete dataset. The detection of such gaps provides valuable information that would be lost if we just used a line diagram as in figure B.1.

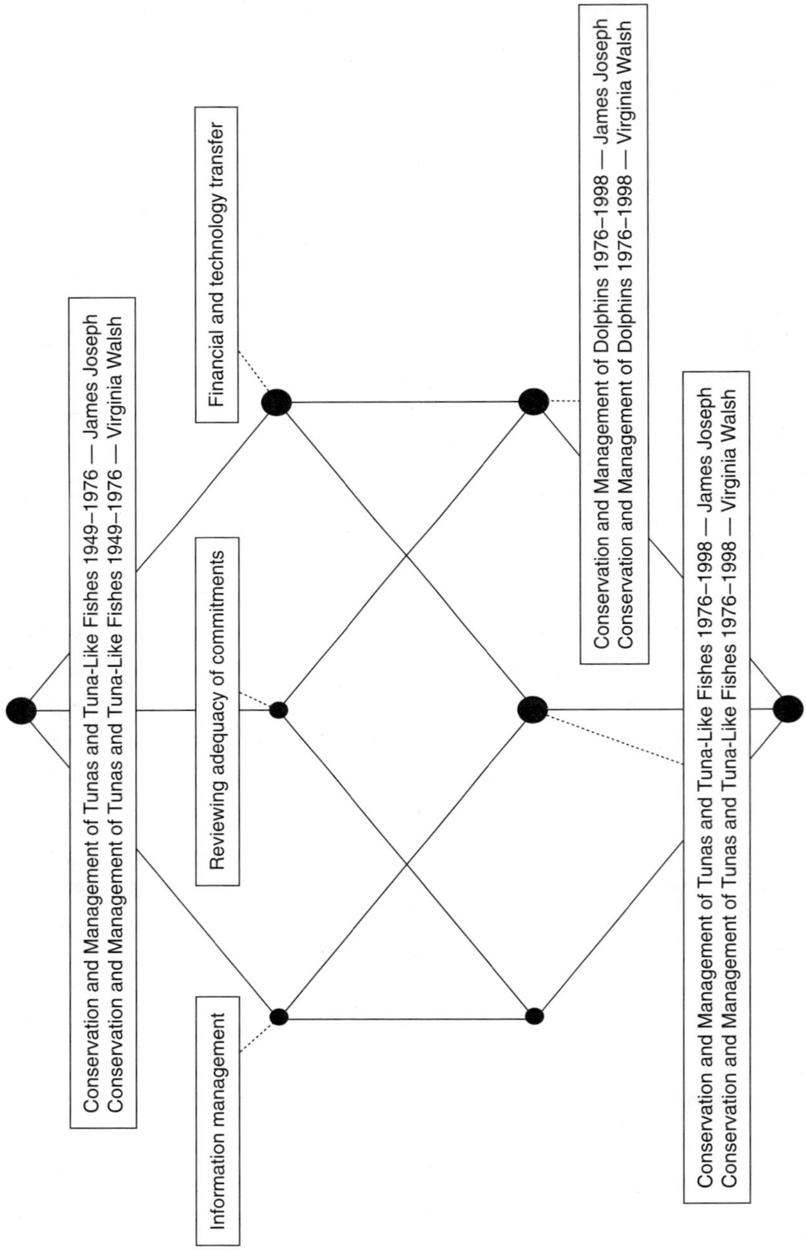


**Figure B.3**
Line diagram in the IRD TOSCANA system

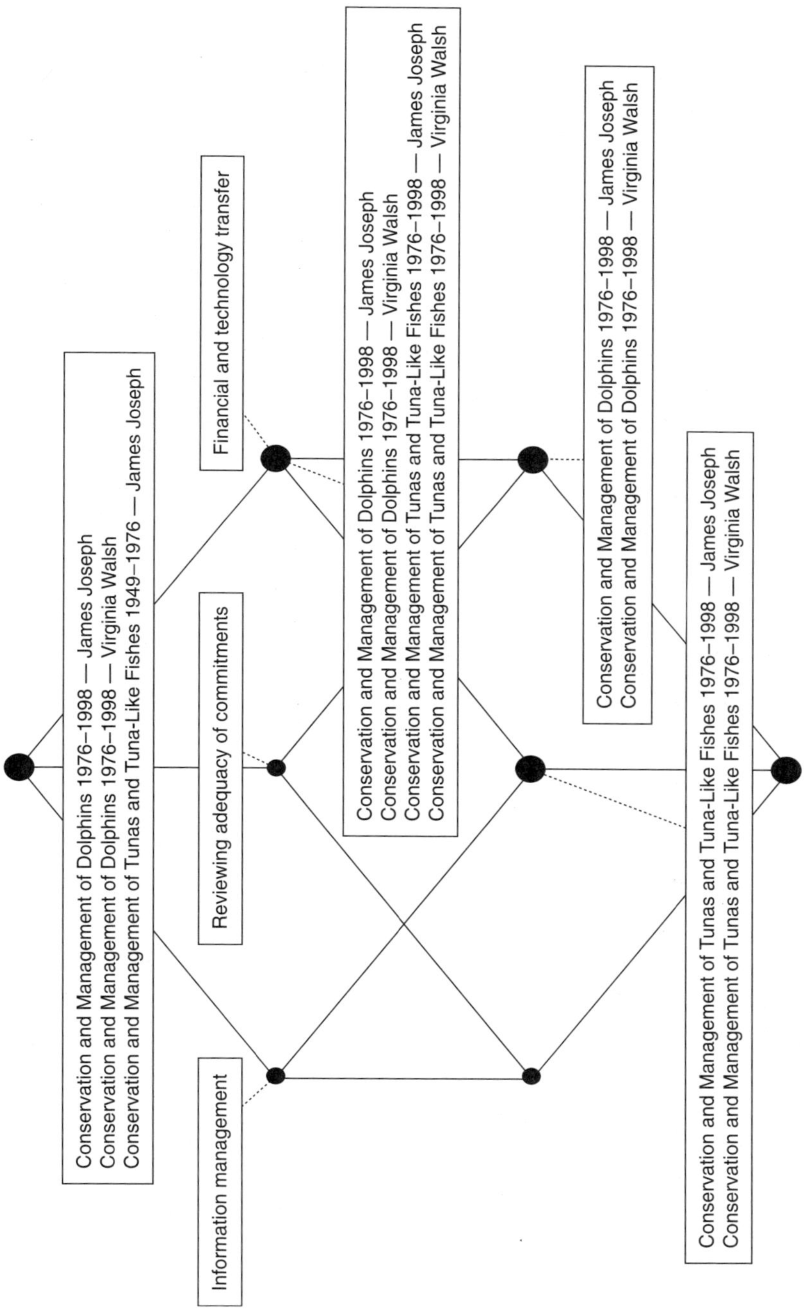


**Figure B.4**
Line diagram, labeling option: show all matches

### B.3.2 Nesting

Using the technique of nesting, we can combine several themes in one diagrammatic view. Chapter 4—"Decision Rules, Compliance Mechanisms, and the Effectiveness of Regimes"—analyzes the connection between compliance mechanisms of a regime component and the regime's effectiveness as measured by the variables on goal fulfillment and problem solving. This analysis seeks to extract correlations between two variables. This is a typical question where a nested line diagram created with the IRD-TOSCANA system gives a good overview of the empirical evidence at hand. As an example, I will consider diagrams for the variables COMPLIANCE_APPROACH from form RA 49 and GOALS_FULFILL from form RC 10. Figure B.5 shows a nested diagram for these themes, "RA 49: Compliance Procedures: Enforcement or Management Approach" and "RC 10: Goal fulfillment." We will see how to read it and what conclusions can be drawn.[1]

Now we can concentrate on what we can deduce regarding the correlation between the two variables. If we look inside the big node labeled "Management Approach," we see that the majority of regime components report a good rate of goal fulfillment. By contrast, we see that in the opposite node labeled "Enforcement Approach" we have the same proportion of regime components having a high success rate (more that 75 percent of goals fulfilled) and having a low success rate (less than 25 percent of goals fulfilled). Therefore, we can use this diagram as evidence in support of the hypothesis that a management approach does better in supporting goal fulfillment than an enforcement approach does. Nonetheless, the diagram forces us to recognize that there are only ten regime components approaching compliance with an enforcement approach. It follows that our confidence in the available evidence should be somewhat circumscribed.

### B.3.3 Zooming

As chapter 6 makes clear, it is often interesting to look at certain subsets of the universe of regime components when considering special topics. Zooming makes it possible to examine relevant subsets. For instance, we can ask: "How are the regime components with a high rate of goal

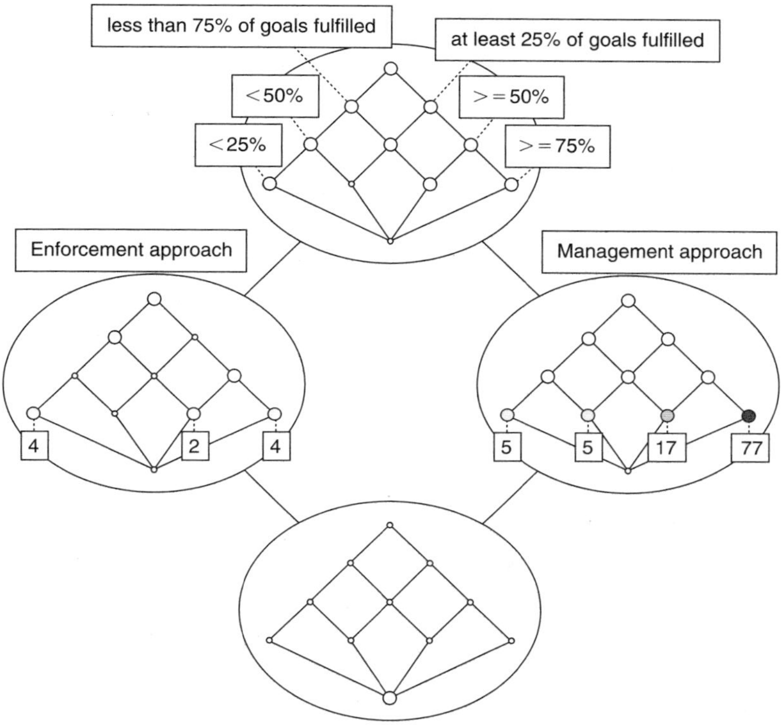


**Figure B.5**
Nested diagram. A nested diagram consists of an outer diagram and an inner diagram. The nodes of the outer diagram—here for the theme "RA49: Compliance Procedures: Enforcement or Management Approach"—are represented by big circles where each big circle contains a copy of the diagram for the inner theme, here "RC10: Goal Fulfillment." To save space the labeling for the inner diagram is only shown in the top node. So the objects that would have appeared at the nodes of the outer diagram are distributed about the inner one. In our example, we see that among the regime components that prefer a management approach, seventy-seven have a rate of goal fulfillment higher than 75 percent. The nodes are shaded according to their degree of fulfillment.

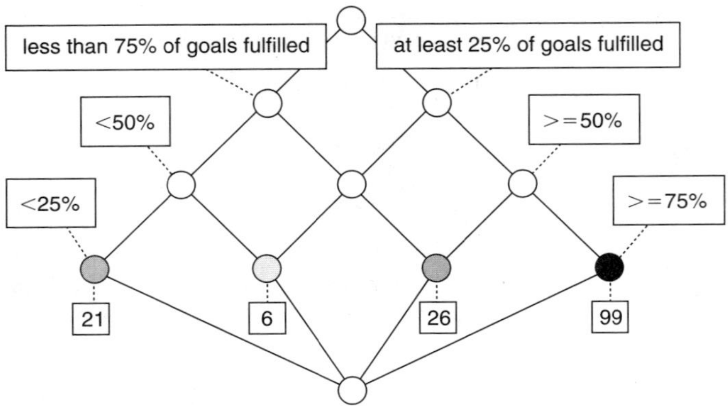


**Figure B.6**
Diagram for the theme "Goal Fulfillment"

fulfillment distributed between the two compliance approaches?" To answer this question, we need only make a few mouse-clicks in the IRD-TOSCANA system. We set the depth of nesting to "Flat Diagram." Then, we select the theme "RC 10: Goal Fulfillment." This produces the diagram in figure B.6.

Now, we select the diagram named "RA 49: Compliance Procedures: Enforcement or Management Approach." It will appear in a list of selected diagrams. But since the depth of nesting is set to one, the diagram for the second theme is not displayed. A double-click on the concept node of the concept for ">=75%" yields the visualization in figure B.7.

This diagram shows a subset of the ninety-nine regime components with goal fulfillment equal to or higher than 75 percent.[2] The zooming process has filtered the object set, considering only the objects for inclusion in the next diagram that fulfill all attributes contained in the intent of the selected concept.

From figure B.7, we can—knowing that we zoomed in to the concept for ">=75%" in the diagram for "Goal Fulfillment"—extract an interesting hypothesis. Thus, we could answer our question in the following way: a high level of goal fulfillment is associated with a high probability of a management approach regarding compliance.

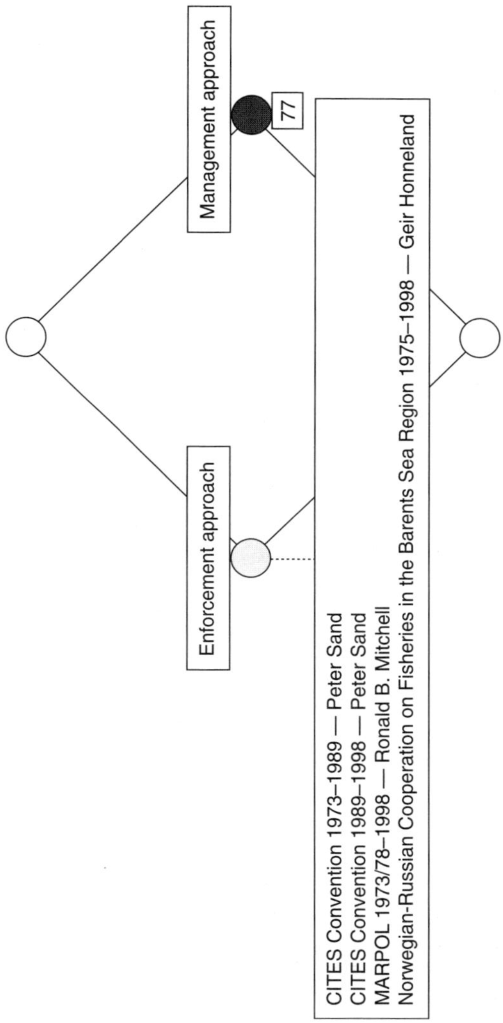


**Figure B.7**
Diagram for the theme "Enforcement"

It is important to be aware that this rule is derived only from data that are obtained from observations of the past. This means that there is no guarantee that this rule will hold in the future. But moving toward a theory that can prescribe designs for international regimes that lead to higher success rates requires that relevant domain data and their inherent structure be taken into consideration.

## B.4 Development Process of the TOSCANA System for the IRD

In this section, I report on the development process of the TOSCANA system for the IRD. This involves general issues that would arise under any circumstances when constructing a TOSCANA system, as well as special issues applying only to the TOSCANA system for the IRD.

Figure B.8 sums up the development process of a TOSCANA system broken down into six steps in which different roles of experts are needed.

The first step in building a TOSCANA system, which has implications for all the following steps, is to identify the objects of discourse. This seems trivial but turned out to be a very important decision for the

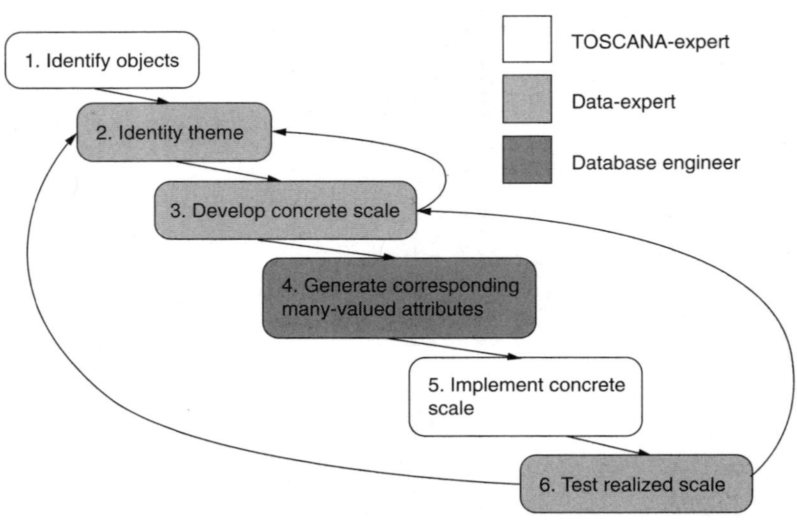


**Figure B.8**
Development process of a TOSCANA system

IRD-TOSCANA system. First, we considered regimes as objects, but this disabled the analysis of internal regime development.[3] Choosing a finer granularity and picking regime components as objects enabled us to analyze internal regime development by looking at different pieces of information for different parts of the whole regime divided by *watersheds*. But there remained a final problem. The IRD contains—at least for most of the coded regime components—information for every regime component supplied by two independent coders. The solution to this problem was to recognize an object as a regime component together with a coder. This resulted in doubling the number of objects from 92 to 184, as a result of splitting each component into the two coder parts. But since the number of objects was still small enough for efficient computation and comfortable analysis for the end user, we incorporated the doubling by appending the coder to each component name. This was an important step since it allows users of the system to compare attributes of a regime component coded by different coders. A user can immediately recognize situations in which coders differ in their assessment of specific attributes of the regime component because they do not appear in the labeling of the same node. For instance, in figure B.4 we can see immediately that Virginia Walsh and James Joseph were in agreement in judging the regime component *Conservation and Management of Dolphins 1976–1998*. If one of them appeared alone in a label, we would have known immediately that the other one differed in assessing the attribute, and we can find that coder's estimation when we locate her or his label in the diagram.

Turning to steps 2 to 6 in figure B.7, we can see that they are iterative in the sense that the backward arrows indicate optional backtracking possibilities. For instance, if a scale is regarded as inappropriate in step 6, we can dismiss it and go back to step 2.

Step 2 addresses the identification of *themes*. A theme is understood as a coherent part of the underlying domain. In the IRD system, themes often stem from forms of the IRD Data Protocol (Breitmeier et al. 1996a), since the protocol is broken down into exactly such pieces by the forms. An example of a theme is "Programmatic activities: Management," which is a part of the form RA 16 (Breitmeier et al. 1996a, 86). Here it was necessary to break the form down into three themes, since every attribute combination is possible and the resulting diagram would have become too large.

Step 3, the development of conceptual scales, involves the interpretation of the data for individual themes. For the theme "Programmatic activities: Management," for instance, we argued that the three programmatic activities may combine in an arbitrary fashion; this led to the scale shown in figure B.4. This scale demonstrates that we did not hypothesize any linkages among the attributes a priori. Still, in the TOSCANA system, links can be found as described in section B.2 when the scale is realized with "real" objects from the database. So the development of conceptual scales frequently leads to discussions about the embedded theory for a small part of the data to be analyzed. This makes it important for successful development to have both TOSCANA experts and domain experts present.

Step 4, the generation of the corresponding attributes for the scales, addresses the possible need for data transformation to fulfill the requirements of TOSCANA. Clearly, in the case of the IRD-TOSCANA system, lots of transformations were necessary, since the objects as displayed by the IRD-TOSCANA system are not present in the tables of the IRD database. We transformed the data using SQL[4] queries, which extracted and combined the data into new tables that are accessible via the TOSCANA system.

For the implementation of the conceptual scales, we used the additional software *Elba*, which is part of the ToscanaJ package. This program supports the geometric definition of the actual diagrams and the formulation of the SQL queries that are used by TOSCANA to retrieve the objects from the database.

The last step of our sequence involves testing the developed scale using TOSCANA and checking to see if the realized scale is easily readable, displays correctly, and allows users to extract valuable information from it.

These considerations make it obvious that the development process of a TOSCANA system requires discussion regarding the structure of the data and therefore often produces deeper insights into the data, and especially into possible influences on the interpretation that are dependent on the data structure.

## B.5 Special Features of the System

Currently, the TOSCANA system for the IRD contains more than two hundred conceptual scales that make it possible to browse a wide range of topics in the underlying database.

In addition to the functions mentioned already, the IRD-TOSCANA system supports several detail views. A right-click on a regime component opens a popup menu that allows viewers to see different properties of the selected object.

### B.5.1 Problems Detail View

If we choose the problems property for a regime component, we get the sort of popup illustrated in box B.1. As the codebook makes clear, coders may assign up to three problems to a regime.

### B.5.2 Goals Detail View

If we decide to inform ourselves about the goals a regime component pursues, we can use as a popup the one displayed in box B.2. This regime component identifies three goals. The numbers following each goal refer to the level of goal fulfillment the coder has attributed to that component.

### B.5.3 Rules Detail View

We can even access from the TOSCANA-front end a list of rules that regime members are supposed to follow. Box B.3 provides an illustration drawn from the Danube River regime.

**Box B.1**

**Antarctic Treaty 1989/91–1998—MJ Peterson**

*Problems:*
Growth of Interest in Managing Exploitation of Resources in and around Antarctica
Jurisdictional Differences/Conflicts Between Claimants and Major Non-claimant States
Jurisdictional Differences/Conflicts About Overlapping Claims on the Part of Claimant States

**Box B.2**

## Antarctic Treaty 1989/91–1998—MJ Peterson

*Goals:*
"2" indicates "goal fulfilled", "1" indicates "goal not fulfilled" and "0" indicates "do not know".
Reservation of Antarctica to peaceful activity (to ensure that Antarctica is used for peaceful purposes only) 2
To promote international scientific cooperation in Antarctica 2
Avoidance of international conflict about Antarctic territory 2

**Box B.3**

## Danube River Protection 1991–1994 —Andy Garner

*Rules:*
Development of an Accident Emergency Warning System for the river basin to increase the safety, to protect the environment and to enable national authorities to protect water users against accidental pollution and other emergency situations by providing early information on transboundary water pollution incidents for the affected riparian countries and development of detailed procedures of operation in an International Operations Manual for Principal International Alert Centres (PIACS).

To establish a Monitoring, Laboratory Management and Information Management (MLIM) program with the aim to strengthen national and international capacity to provide reliable information on surface water flows and the quality of waters in the Danube river basin, to improve the comparability of sampling techniques and laboratory analysis, and to develop compatible information management systems for the exchange of information at the international level.

Development of a Transnational Monitoring Network (TNMN) to strengthen the existing network set up by the Bucharest Declaration, and which shall be capable of supporting reliable and consistent trend analysis for concentrations and loads for priority pollutants, to support the assessment of water quality for water use, to assist in the identification of major pollution sources, to include sediment monitoring and bioindicators, to include quality control.

View Item

**United Nations Convention to Combat Desertification 1994-1998 - Elisabeth Corell**

| Program Activities | |
|---|---|
| Scientific monitoring | |
| Compliance monitoring | |
| Review of implementation | x |
| Research | |
| Financial and technology transfer | x |
| Information management | |
| Verification of compliance | x |
| Reviewing adequacy of commitments | |
| Expert advice | x |

Close

**Figure B.9**
Detail view *Programmatic Activities*

### B.5.4 Programmatic Activities Detail View

Users interested in the programmatic activities of a component can get an overview table from the context menu, as illustrated in figure B.9.

## B.6 Conclusion

This appendix demonstrates the high potential of the IRD system for analysts interested in international regimes. Although it takes some time to become familiar with reading the diagrams and using the TOSCANA software, the possibilities for exploring the data, even in a playful way, are immense. Browsing the diagrams using zooming and nesting and selecting detail views when desired can verify or falsify a given hypothesis as well as produce interesting new ones without worrying about the technical details of the database, such as SQL queries, since these details are hidden by the system.

Since ToscanaJ is an open-source software, those interested in using it can download this software free of charge.[5] You may contact any of the members of the IRD team to retrieve the IRD-TOSCANA system and explore the premise underlying this appendix that the IRD-TOSCANA system provides valuable support for those desiring to use the IRD to enhance their understanding of international regimes.

# Notes

## Chapter 2

1. Additional software called TOSCANA is available for use with a subset of the variables included in the IRD (Stumme and Wille 2000). On the application of TOSCANA to the data included in the IRD, see the contribution of Tim Kaiser included in appendix B.

## Chapter 3

1. This chapter employs a conceptual framework similar to that of Zürn 2004, but applies it to a different setting and relies on different empirical applications and a different empirical context.

2. See especially the work of Bothe (1996); Chayes and Chayes (1993, 1995); Haas (1998); Mitchell (1994a); Underdal (1998); Victor, Raustiala, and Skolnikoff (1998); Weiss and Jacobson (1998); Young (1979); Young (1999b, chap. 4).

3. This is not the same as the traditional notion of countermeasures, according to which "the injured state enforced its own rights through self-help" (O'Connell 1995, 2).

4. Intercoder reliability regarding this variable is satisfactory. For more than 80 percent of the regime elements coded by two experts, data regarding compliance on the part of all members differed by no more than one value on the measurement scale.

5. Ramsar COP7/DOC.5, *Report of the Secretary General*, 7th Meeting of the Conference of the Parties to the 1971 Convention on Wetlands, San José, Costa Rica, May 10–18, 1999.

6. The fourth source of noncompliance—the lack of capacity to implement—needs further elaboration. The underlying "challenge" to the rule in this case is somewhat different from an outright questioning of its validity. The practicability rather than the normative validity of the rule is challenged. Hence, the immediate response to these cases of noncompliance is more often a discourse about the

possibility of altering the formulation of the rule to make it more effective, rather than simply scrapping it. In this sense, the rule is challenged less than in the case of cell 3.

7. For other categorizations of approaches explaining compliance see, among others, Haas 1998; Hathaway 2002; Hurd 1999; Weiss and Jacobson 1998; Underdal 1998; Vogel and Kessler 1998.

8. For discussions of underlying strategic settings and appropriate institutional design, see Axelrod and Keohane 1985; Martin 1993; Mitchell and Keilbach 2001; Scharpf 1997; Stein 1983; Zürn 1992. The most important determinant of cooperation and compliance in the view of rational institutionalism is the so-called situation structure or, in other words, the interest constellation that gives rise to the collective-action problem (Oye 1986; Zürn 1992; Martin 1993). See Hasenclever et al. 1997 for an overview. In our study, we aim to keep the situation structure constant by comparing similar regulations across levels.

9. The major work on this theme is H. L. A. Hart, *The Concept of Law* (1961). The classic formulation of Oliver Wendell Holmes's (1897) concept of legal realism—"Law is what courts do"—should also be noted.

10. See also Goldstein and Martin 2000. Legalization, following our terminology, describes a double process of juridification and internalization. What Abbott et al. (2000) define as legalization is in this sense very close to our concept of juridification.

11. On this pair of terms, see Elster 1992, 1998; Gehring 1996. Risse 2000 provides a treatment of communicative action in world politics.

12. Keohane, Moravcsik, and Slaughter (2000) discuss delegation—the major component of our concept of juridification—in such a way as to include what we treat as internalization. Besides independence, defined as "the extent to which formal legal arrangements ensure that adjudication can be rendered impartially," they also consider "access" (the "ease with which parties other than states can influence the tribunal's agenda") and "embeddedness," which "denotes the extent to which dispute resolution decisions can be implemented without governments having to take actions to do so" as part of "delegation."

13. While the principle of direct effect is unique to the EU, some domestic legal systems allow for the enforcement of international law without prior implementation by the national legislature. O'Connell (1995, 5) cites the *Paquette Habana* as the most famous case in the United States: "In *Paquette Habana* U.S. Navy ships arrested Cuban fishing vessels during the Spanish-American War. The Navy wanted to sell the vessels as prizes of war. The United States Supreme Court held that under international law, fishing vessels cannot be captured as prizes of war."

14. There is also one case in which an NGO performs the role of the secretariat.

15. We should add that the causal role coders assigned to the level of independence of the secretariat was marginal.

16. On the other hand, the decision rules provided for in the treaties seem far less important for compliance issues. In any case, decision rules that forgo the traditional consensus principle of international relations do not result in higher compliance rates. According to the causal attributions of coders, it appears to be the other way around.

17. From this perspective, deliberation is mainly a matter of communicative action in dyadic settings and has little to do with legal reasoning in triadic settings, which is bound up with the legalization perspective.

18. This emphasis on legal *process* is evidence of significant similarities between this perspective on compliance and the approach of the so-called New Haven School (McDougal and Associates 1960).

## Chapter 4

1. Unless otherwise noted, full texts of the conventions and treaties referred to in this section are reprinted in Weiss, Szasz, and Magraw 1992.

2. The full text of the Convention on the Regulation of Antarctic Mineral Resource Activities is reprinted in *International Legal Materials*, 27 (July 1988): 859–900.

3. Implicitly, at least, this line of thinking assumes that it is possible to construct a social-welfare function.

4. Of course, such mechanisms will be the norm when the governments of member states endeavor to implement the terms of international regimes with regard to the actions of those operating under their jurisdiction.

5. The universes of cases in the two queries are overlapping but not identical.

6. The alert reader will notice some discrepancy between our discussion of capacity building in this chapter and in chapter 3. We will return to this matter in chapter 6.

## Chapter 5

1. On the role of knowledge as a factor that influences the formation and effectiveness of regimes, see Adler and Haas 1992; E. Haas 1990; P. Haas 1990. On the impact of international institutions on social learning in individual countries consult Social Learning Group 2001a, 2001b.

2. See Article II of the Great Lakes Water Quality Agreement of 1978, with Annexes and Terms of Reference, between the United States and Canada signed at Ottawa November 22, 1978, and the Phosphorus Load Reduction Supplement signed October 16, 1983, as amended by a Protocol signed November 18, 1987.

3. Scientific assessment focuses on the integration of existing knowledge. Research, by contrast, features efforts to generate new knowledge.

4. Rational-choice models have begun to address the consequences of relaxing the assumption of fixed preferences. Interest in two-level games is rising among those who work on such models. But the development of formal models that take into account both international and domestic levels of policymaking is fraught with problems (Putnam 1988; Zangl 1999b).

## Appendix B

1. I only consider cases of nesting in which two diagrams are involved, as supported in ToscanaJ. Theoretically any number of nestings is possible.

2. There are only eighty-one components in the diagram since eighteen coders made no judgment on the compliance approach.

3. The development team consisted of Prof. Dr. Rudolf Wille, Dr. Helmut Breitmeier, Nina Bressel, and Dipl.-Math. Tim Kaiser.

4. SQL = Structured Query Language.

5. http://toscanaj.sourceforge.net/.

# References

Abbott, Kenneth W., Robert O. Keohane, Andrew Moravcsik, Anne-Marie Slaughter, and Duncan Snidal. 2000. "The Concept of Legalization." *International Organization* 54(3): 401–419.

Adler, Emanuel, and Peter M. Haas. 1992. "Conclusion: Epistemic Communities, World Order, and the Creation of a Reflective Research Program." *International Organization* 46(1): 367–390.

Allott, Philip. 1999. "The Concept of International Law." *European Journal of International Law* 10(1): 31–50.

Alter, Karen J. 2001. *Establishing the Supremacy of European Law: The Making of an International Rule of Law in Europe*. Oxford: Oxford University Press.

Axelrod, Robert. 1984. *The Evolution of Cooperation*. New York: Basic Books.

Axelrod, Robert, and Robert O. Keohane. 1985. "Achieving Cooperation under Anarchy: Strategies and Institutions." *World Politics* 38(1): 226–254.

Barrett, Scott. 2003. *Environment and Statecraft*. New York: Oxford University Press.

Beisheim, Marianne. 2004. *Fit für Global Governance? Transnationale Interessengruppenaktivitäten als Demokratisierungspotential—am Beispiel Klimapolitik*. Opladen: Leske und Budrich.

Benedick, Richard Elliott. 1991. *Ozone Diplomacy: New Directions in Safeguarding the Planet*. Cambridge, MA: Harvard University Press.

Bernauer, Thomas. 1995. "The Effect of International Environmental Institutions: How We Might Learn More." *International Organization* 49(2): 351–377.

Bernauer, Thomas. 1996. "Protecting the River Rhine against Chloride Pollution." In Robert O. Keohane and Marc A. Levy, eds., *Institutions for Environmental Aid*, 201–232. Cambridge, MA: MIT Press.

Bothe, Michael. 1996. "The Evaluation of Enforcement Mechanisms in International Environmental Law." In Rüdiger Wolfram, ed., *Enforcement Environmental Standards: Economic Mechanisms as Viable Means?*, 13–38. Berlin: Springer.

Breit, Heiko, Anita Engels, Timothy Moss, and Markus Troja, eds. 2003. *How Institutions Change: Perspectives on Social Learning in Global and Local Environmental Contexts*. Opladen: Leske und Budrich.

Breitmeier, Helmut, Marc A. Levy, Oran R. Young, and Michael Zürn. 1996a. "International Regimes Database (IRD): Data Protocol." IIASA Working Paper WP-96-154. Laxenburg: International Institute for Applied Systems Analysis.

Breitmeier, Helmut, Marc A. Levy, Oran R. Young, and Michael Zürn. 1996b. "The International Regimes Database as a Tool for the Study of International Cooperation." IIASA Working Paper WP-96-160. Laxenburg: International Institute for Applied Systems Analysis.

Breitmeier, Helmut, and Volker Rittberger. 2000. "Environmental NGOs in an Emerging Global Civil Society." In Pamela S. Chasek, ed., *The Global Environment in the Twenty-First Century: Prospects for International Cooperation*, 130–163. Tokyo: United Nations University Press.

Buchanam, James M., and Gordon Tullock. 1962. *The Calculus of Consent*. Ann Arbor: University of Michigan Press.

Bull, Hedley. 1977. *The Anarchical Society: A Study of Order in World Politics*. New York: Columbia University Press.

Chayes, Abram, and Antonia Handler Chayes. 1993. "On Compliance." *International Organization* 47(2): 175–205.

Chayes, Abram, and Antonia Handler Chayes. 1995. *The New Sovereignty: Compliance with International Regulatory Agreements*. Cambridge, MA: Harvard University Press.

Checkel, Jeffrey. 2001. "Why Comply? Social Learning and European Identity Change." *International Organization* 55: 553–588.

Cutler, A. Claire, Virginia Haufler, and Tony Porter, eds. 1999. *Private Authority in International Affairs*. Albany: SUNY Press.

Downs, George W., David M. Rocke, and Peter N. Barsoom. 1996. "Is the Good News about Compliance Good News about Cooperation?" *International Organization* 50: 379–406.

Dworkin, Ronald. 1986. *Law's Empire*. Cambridge, MA: Harvard University Press.

Dworkin, Ronald. 1991. *Taking Law Seriously: New Impressions with a Reply to Critics*. London: Duckworth.

Ebbin, Syma, ed. 2004. Special issue on Knowledge and Institutions. *International Environmental Agreements*, 4(2).

Ehrmann, Markus. 2002. "Procedures of Compliance Control in International Environmental Treaties." *Colorado Journal of International Environmental Law and Policy* 13(2): 377–443.

Elster, Jon. 1992. "Arguing and Bargaining in the Federal Convention and the *Assemblée Constituante*." In Raino Malnes and Arild Underdal, eds., *Rationality and Institutions: Essays in Honour of Knut Midgaard*, 13–50. Oslo: Scandinavian University Press.

Elster, Jon. 1998. *Deliberative Democracy*. Cambridge: Cambridge University Press.

Fearon, James D. 1996. "Causes and Counterfactuals in Social Science: Exploring an Analogy between Cellular Automata and Historical Processes." In Philip Tetlock and Aaron Belkin, eds., *Counterfactual Thought Experiments in World Politics: Logical, Methodological, and Psychological Perspectives*. Princeton, NJ: Princeton University Press.

Franck, Thomas M. 1990. *The Power of Legitimacy among Nations*. New York: Oxford University Press.

Gale, Fred P. 1998. *The Tropical Timber Trade Regime*. Basingstoke, UK: Macmillan.

Ganter, Bernhard, and Rudolf Wille. 1989. "Conceptual Scaling." In Frank Roberts, ed., *Applications of Combinatorics and Graph Theory to the Biological and Social Sciences*, 139–167. Heidelberg: Springer.

Ganter, Bernhard, and Rudolf Wille. 1999. *Formal Concept Analysis: Mathematical Foundations*. Heidelberg: Springer.

Gehring, Thomas. 1994. *Dynamic International Regimes: Institutions for International Environmental Governance*. Frankfurt: Lang.

Gehring, Thomas. 1996. "Arguing und Bargaining in internationalen Verhandlungen: Überlegungen am Beispiel des Ozonschutzregimes." In Volker von Prittwitz, ed., *Verhandeln und Argumentieren: Dialog, Interessen und Macht in der Umweltpolitik*, 207–238. Opladen: Leske und Budrich.

Goldstein, Judith, and Lisa L. Martin. 2000. "Legalization, Trade Liberalization, and Domestic Politics: A Cautionary Note." *International Organization* 54(3): 603–632.

Greene, Owen. 1998. "Implementation Review and the Baltic Sea Regime." In David G. Victor, Kal Raustiala, and Eugene B. Skolnikoff, eds., *The Implementation and Effectiveness of International Environmental Commitments: Theory and Practice*, 177–220. Cambridge, MA: MIT Press.

Großkopf, Anja, und Gisela Harras. 2000. "Begriffliche Erkundung semantischer Strukturen von Sprechaktverben." In Gerd Stumme and Rudolf Wille, eds., *Begriffliche Wissensverarbeitung. Methoden und Anwendungen*, 273–295. Springer: Berlin.

Haas, Ernst B. 1990. *When Knowledge Is Power: Three Models of Change in International Organizations*. Berkeley: University of California Press.

Haas, Peter M. 1990. *Saving the Mediterranean: The Politics of International Environmental Cooperation*. New York: Columbia University Press.

Haas, Peter M. 1992. "Introduction: Epistemic Communities and International Policy Coordination." *International Organization* 46(1): 1–35.

Haas, Peter M. 1998. "Compliance with EU Directives: Insights from International Relations and Comparative Politics." *Journal of European Public Policy* 5(1): 17–37.

Haas, Peter M., Robert O. Keohane, and Marc A. Levy, eds. 1993. *Institutions for the Earth: Sources of Effective International Environmental Protection*. Cambridge, MA: MIT Press.

Habermas, Jürgen. 1994. *Faktizität und Geltung. Beiträge zur Diskurstheorie des Rechts und des demokratischen Rechtsstaats*. Frankfurt: Suhrkamp.

Haggard, Stephan, and Beth A. Simmons. 1987. "Theories of International Regimes." *International Organization* 41: 491–517.

Hart, H. L. A. 1961. *The Concept of Law*. Oxford: Clarendon Press.

Hasenclever, Andreas, Peter Mayer, and Volker Rittberger. 1997. *Theories of International Regimes*. Cambridge: Cambridge University Press.

Hasenclever, Andreas, Peter Mayer, and Volker Rittberger. 2004. "Does Regime Robustness Require a Fair Distribution of the Gains from Cooperation?" In Arild Underdal and Oran R. Young, eds., *Regime Consequences*, 183–216. Dordrecht: Kluwer Academic Publishers.

Hathaway, Oona. 2002. "Do Human Rights Treaties Make a Difference?" *Yale Law Journal* 111(8): 1935–2042.

Helm, Carsten, and Detlef Sprinz. 1999. *Measuring the Effectiveness of International Environmental Regimes*. PIK Report No. 52. Postdam: Postdam Institute for Climate Impact Research.

Henkin, Louis. 1979. *How Nations Behave: Law and Foreign Policy*. 2nd ed. New York: Columbia University Press.

Hereth, Joachim, Gerd Stumme, Rudolf Wille, and Uta Wille. 2000. "Conceptual Knowledge Discovery in Data Analysis." In Bernhard Ganter and Guy W. Mineau, eds., *Conceptual Structures: Logical, Linguistic, and Computational Issues*, 421–437. Heidelberg: Springer.

Hereth, Joachim, Gerd Stumme, Uta Wille, and Rudolf Wille. 2003. "Conceptual Knowledge Processing—A Human-Centered Approach." *Journal on Applied Artificial Intelligence* 17: 3.

Hoffmann, Matthew J. 2005. *Ozone Depletion and Climate Change*. Albany: SUNY Press.

Holmes, Oliver Wendell. 1897. "The Path of the Law." *Harvard Law Review* 10: 457–467.

Hurd, Ian. 1999. "Legitimacy and Authority in International Politics." *International Organization* 53(2): 379–408.

Hurrell, Andrew. 1993. "International Society and the Study of Regimes: A Reflective Approach." In Volker Rittberger, ed., *Regime Theory and International Relations*, 49–72. Oxford: Clarendon Press.

Joerges, Christian. 2000. "Transnationale deliberative Demokratie oder deliberativer Surpranationalismus? Anmerkungen zur Konzeptualisierung legitimen

Regierens jenseits des Nationalstaats bei Rainer Schmalz-Bruns." *Zeitschrift für Internationale Beziehungen* 7(1): 145–161.

Jönsson, Christer, and Jonas Tallberg. 1998. "Compliance and Post-Agreement Bargaining." In *European Journal of International Relations* 4(4): 371–408.

Joyner, Christopher C. 1998. *Governing the Frozen Commons: The Antarctic Regime and Environmental Protection*. Columbia: University of South Carolina Press.

Kahneman, Daniel. 2003. "Maps of Bounded Rationality: Psychology for Behavioral Economics." *American Economic Review* 93: 1449–1475.

Kaspar, Martin. 1999. *Erfolgsfaktoren regionaler Umweltprogramme in Mittel- und Osteuropa*. Wien: WUV.

Keane, John. 2003. *Global Civil Society*. Cambridge: Cambridge University Press.

Keck, Margaret, and Kathryn Sikkink. 1998. *Activists beyond Borders: Transnational Advocacy Networks in International Politics*. Ithaca, NY: Cornell University Press.

Keene, Edward. 2002. *Beyond the Anarchical Society: Grotius, Colonialism and Order in World Politics*. Cambridge: Cambridge University Press.

Keohane, Robert O. 1984. *After Hegemony: Collaboration and Discord in the World Political Economy*. Princeton, NJ: Princeton University Press.

Keohane, Robert O. 1988. "International Cooperation: Two Approaches." *International Studies Quarterly* 32(4): 379–396.

Keohane, Robert O., and Lisa L. Martin. 1995. "The Promise of Institutionalist Theory." *International Security* 20(1): 39–51.

Keohane, Robert O., Andrew Moravcsik, and Anne-Marie Slaughter. 2000. "Legalized Dispute Resolution: Interstate and Transnational." *International Organization* 54(3): 457–488.

Khagram, Sanjeev, James V. Riker, and Kathryn Sikkink, eds. 2002. *Restructuring World Politics: Transnational Social Movements, Networks, and Norms*. Minneapolis: University of Minnesota Press.

King, Gary, Robert O. Keohane, and Sydney Verba. 1994. *Designing Social Inquiry: Scientific Inference in Qualitative Research*. Princeton, NJ: Princeton University Press.

Koh, Harold Hongju. 1997. "Why Do Nations Obey International Law?" *Yale Law Journal* 106(8): 2599–2659.

Kohler-Koch, Beate, and Frank Vogt. 2000. "Normen und regelgeleitete internationale Kooperationen." In Gerd Stumme and Rudolf Wille, eds., *Begriffliche Wissensverarbeitung. Methoden und Anwendungen*, 239–253. Springer: Berlin.

Kollewe, Wolfgang, Martin Skorsky, Frank Vogt, and Rudolf Wille. 1994. "TOSCANA—ein Werkzeug zur begrifflichen Analyse und Erkundung von Daten." In Rudolf Wille and Monika Zickwolff, eds., *Begriffliche*

*Wissensverarbeitung—Grundfragen und Aufgaben*, 267–288. Mannheim: B.I.-Wissenschaftsverlag.

Koremenos, Barbara, Charles Lipson, and Duncan Snidal. 2001. "The Rational Design of International Institutions." *International Organization* 55: 761–799.

Krasner, Stephen D., ed. 1983a. *International Regimes*. Ithaca, NY: Cornell University Press.

Krasner, Stephen D. 1983b. "Structural Causes and Regime Consequences: Regimes as Intervening Variables." In Stephen D. Krasner, ed., *International Regimes*, 1–21. Ithaca, NY: Cornell University Press.

Kratochwil, Friedrich. 1989. *Rules, Norms, and Decisions: On the Conditions of Practical and Legal Reasoning in International Relations and Domestic Affairs*. Cambridge: Cambridge University Press.

Lefeber, René. 2002. "From The Hague to Bonn to Marrakesh and Beyond: A Negotiating History of the Compliance Regime under the Kyoto Protocol." In *Hague Yearbook of International Law* 2001: 25–54.

Levy, Marc A. 1993. "European Acid Rain: The Power of Tote-Board Diplomacy." In Peter M. Haas, Robert O. Keohane, and Marc A. Levy, eds., *Institutions for the Earth: Sources of Effective International Environmental Protection*, 75–132. Cambridge, MA: MIT Press.

Levy, Marc A. 1995. "International Cooperation to Combat Acid Rain." In Helge Ole Bergesen, Georg Parmann, and Oystein B. Thommessen, eds., *Green Globe Yearbook of International Co-operation on Environment and Development 1995*, 59–68. Oxford: Oxford University Press.

Levy, Marc A., Oran R. Young, and Michael Zürn. 1995. "The Study of International Regimes." *European Journal of International Relations* 1: 267–330.

Litfin, Karen T. 1994. *Ozone Discourses: Science and Politics in Global Environmental Cooperation*. New York: Columbia University Press.

Lomborg, Bjørn. 2001. *The Skeptical Environmentalist*. Cambridge: Cambridge University Press.

Marauhn, Thilo. 1996. "Towards a Procedural Law of Compliance Control in International Environmental Relations." *Zeitschrift für ausländisches öffentliches Recht und Völkerrecht* (ZaöRV) 56: 696–731.

March, James G., and Johan P. Olsen. 1989. *Rediscovering Institutions: The Organizational Basis of Politics*. New York: Free Press.

March, James G., and Johan P. Olsen. 1998. "The Institutional Dynamics of International Political Orders." *International Organization* 52: 943–969.

Martin, Lisa L. 1993. "The Rational State Choice of Multilateralism." In John Gerard Ruggie, ed., *Multilateralism Matters: The Theory and Practice of an International Form*, 91–121. New York: Columbia University Press.

McDougal Myres and Associates. 1960. *Studies in World Public Order*. New Haven: Yale University Press.

Mearsheimer, John J. 1994–1995. "The False Promise of International Institutions." *International Security* 19: 5–49.

Miles, Edward L., Arild Underdal, Steinar Andresen, Jørgen Wettestad, Jon Birger Skjaerseth, and Elaine M. Carlin. 2002. *Environmental Regime Effectiveness: Confronting Theory with Evidence*. Cambridge, MA: MIT Press.

Mitchell, Ronald B. 1994a. *Intentional Oil Pollution at Sea: Environmental Policy and Treaty Compliance*. Cambridge, MA: MIT Press.

Mitchell, Ronald B. 1994b. "Regime Design Matters: Intentional Oil Pollution and Treaty Compliance." *International Organization* 48: 425–458.

Mitchell, Ronald B., and Patricia M. Keilbach. 2001. "Situation Structure and Institutional Design: Reciprocity, Coercion, and Exchange." *International Organization* 55(4): 891–917.

Müller, Harald. 1994. "Internationale Beziehungen als Kommunikatives Handeln zur Kritik der Utilitaristischen Handlungstheorien." *Zeitschrift fur Internationale Beziehungen* 1: 15–44.

Neyer, Jürgen. 2004. *Postnationale politische Herrschaft: Vergesellschaftung und Verrechtlichung jenseits des Staates*. Baden-Baden: Nomos.

North, Douglass C. 1990. *Institutions, Institutional Change and Economic Performance*. Cambridge: Cambridge University Press.

Oberthür, Sebastian. 2001. "Linkages between the Montreal and Kyoto Protocols: Enhancing Synergies between Protecting the Ozone Layer and the Global Climate." *International Environmental Agreements: Politics, Law and Economics* 1(3): 357–377.

Oberthür, Sebastian, and Thomas Gehring, eds. 2006. *Institutional Interaction—How to Prevent Conflicts and Enhance Synergies between International and European Environmental Institutions*. Cambridge, MA: MIT Press.

Oberthür, Sebastian, and Hermann Ott. 1999. *The Kyoto-Protocol: International Climate Policy for the 21st Century*. Berlin: Springer-Verlag.

O'Connell, Mary Ellen. 1995. "Enforcement and the Success of International Environmental Law." *Indiana Journal of Global Legal Studies* 3(1): 47–64.

Ostrom, Elinor. 1990. *Governing the Commons: The Evolution of Institutions for Collective Action*. Cambridge: Cambridge University Press.

Oye, Kenneth A. 1986. "Explaining Cooperation under Anarchy: Hypotheses and Strategies." In Kenneth A. Oye, ed., *Cooperation under Anarchy*, 1–24. Princeton, NJ: Princeton University Press.

Parson, Edward A. 2003. *Protecting the Ozone Layer: Science and Strategy*. New York: Oxford University Press.

Parson, Edward A., and Owen Greene. 1995. "The Complex Chemistry of the International Ozone Agreements." *Environment* 37(2): 16–20, 34–43.

Putnam, Robert P. 1988. "Diplomacy and Domestic Politics: The Logic of Two-Level Games." *International Organization* 42: 427–460.

Raustiala, Kal. 1995. *The Domestication of International Commitments*. Working Paper WP-95-115. Laxenburg: International Institute for Applied Systems Analysis.

Raustiala, Kal, and Anne-Marie Slaughter. 2002. "International Law, International Relations and Compliance." In Walter Carlsnaes, Thomas Risse, and Beth A. Simmons, eds., *Handbook of International Relations*, 538–558. London: Sage.

Regimes Summit. 1991. *Workshop Report: 2 December 1991*. Available from the Institute of Arctic Studies at Dartmouth College.

Reinecke, Wolfgang. 1998. *Global Public Policy*. Washington, DC: Brookings Institution.

*Report of the Secretary General, 7th Meeting of the Conference of the Contracting Parties to the Convention on Wetlands* (Ramsar, Iran, 1971), San José, Costa Rica, 10–18 May 1999. Ramsar COP7 DOC.5X.

Riker, William H. 1962. *A Theory of Political Coalitions*. New Haven: Yale University Press.

Risse, Thomas. 2000. "'Let's Argue!' Communicative Action in World Politics." *International Organization* 54(1): 1–40.

Rittberger, Volker, ed. 1990. *International Regimes in East-West Politics*. London: Pinter.

Rittberger, Volker, ed. 1993. *Regime Theory and International Relations*. Oxford: Clarendon Press.

Roan, Sharon. 1989. *Ozone Crisis: The 15-Year Evolution of a Sudden Global Emergency*. New York: Wiley.

Rock, Tammo, and Rudolf Wille. 2000. "Ein TOSCANA-Erkundungssystem zur Literatursuche." In Gerd Stumme and Rudolf Wille, eds., *Begriffliche Wissensverarbeitung: Methoden und Anwendungen*, 239–253. Springer: Berlin.

Roginko, Alexei. 1998. "Domestic Implementation of Baltic Sea Population Commitments in Russia and the Baltic States." In David G. Victor, Kal Raustiala, and Eugene B. Skolnikoff, eds., *The Implementation and Effectiveness of International Environmental Commitments: Theory and Practice*. Cambridge, MA: MIT Press.

Rosenau, James N., and Ernst-Otto Czempiel, eds. 1992. *Governance without Government: Order and Change in World Politics*. Cambridge: Cambridge University Press.

Ruggie, John Gerard. 1983. "International Regimes, Transaction Costs, and Change: Embedded Liberalism in the Postwar Economic Order." In Stephen D. Krasner, ed., *International Regimes*, 195–231. Ithaca, NY: Cornell University Press.

Rutherford, Malcolm. 1994. *Institutions in Economics: The Old and the New Institutionalism*. Cambridge: Cambridge University Press.

Sampson, Gary P., and W. Bradnee Chambers, eds. 2002. *Trade, Environment, and the Millennium*. 2nd ed. Tokyo: United Nations University Press.

Sand, Peter H. 1997. "Commodity or Taboo? International Regulation of Trade in Endangered Species." In Helge Ole Bergesen and Georg Parmann, eds., *Green Globe Yearbook*, 19–36. Oxford: Oxford University Press.

Scharpf, Fritz W. 1997. *Games Real Actors Play: Actor-Centered Institutionalism in Policy Research*, 200–205. Boulder: Westview Press.

Scheich, Patrick, Martin Skorsky, Frank Vogt, Cornelia Wachter, and Rudolf Wille. 1993. "Conceptual Data Systems." In Otto Opitz, Berthold Lausen, and Rüdiger Klar, eds., *Information and Classification*, 72–84. Springer: Berlin.

Scott, W. Richard. 1995. *Institutions and Organizations*. Thousand Oaks, CA: Sage.

Selznick, Philip. 1985. "Focusing Organizational Research on Regulation." In Roger G. Noll, ed., *Regulatory Policy and the Social Sciences*, 363–367. Berkeley: University of California Press.

Simmons, Beth A. 2000. "Compliance in International Relations." Paper presented at the 2000 Annual Meeting of APSA, August 31–September 3, 2000, Washington, DC.

Slaughter, Anne-Marie, Andrew S. Tulumello, and Stepan Wood. 1998. "International Law and International Relations Theory: A New Generation of Interdisciplinary Scholarship." *American Journal of International Law* 92(3): 367–397.

Snidal, Duncan. 2002. "Rational Choice and International Relations." In Walter Carlsnaes, Thomas Risse, and Beth A. Simmons, eds., *Handbook of International Relations*, 73–94. London: Sage.

Social Learning Group. 2001a. *Learning to Manage Global Environmental Risks, Volume 1: A Comparative History of Social Responses to Climate Change, Ozone Depletion, and Acid Rain*. Cambridge, MA: MIT Press.

Social Learning Group. 2001b. *Learning to Manage Global Environmental Risks, Volume 2: A Functional Analysis of Social Responses to Climate Change, Ozone Depletion, and Acid Rain*. Cambridge, MA: MIT Press.

Soroos, Marvin S., and Elena N. Nikitina. 1995. "The World Meteorological Organization as a Purveyor of Global Public Goods." In Robert V. Bartlett, Priya A. Kunan, and Mochu Malik, eds., *International Organizations and Environmental Policy*, 69–82. Westport, CT: Greenwood Press.

Speth, James Gustavus. 2004. *Red Sky at Morning*. New York: Oxford University Press.

Stein, Arthur A. 1983. "Coordination and Collaboration: Regimes in an Anarchic World." In Stephen D. Krasner, ed., *International Regimes*, 115–140. Ithaca, NY: Cornell University Press.

Stokke, Olav Schram, and Davor Vidas, eds. 1996. *Governing the Antarctic: The Effectiveness and Legitimacy of the Antarctic Treaty System*. Cambridge: Cambridge University Press.

Stone Sweet, Alec, and James A. Caporaso. 1998. "From Free Trade to Supranational Polity: The European Court and Integration." In Wayne Sandholtz and Alec Stone Sweet, eds., *European Integration and Supranational Governance*, 92–133. Oxford: Oxford University Press.

Strange, Susan. 1983. "*Cave! hic dragones:* A Critique of Regime Theory." In Stephen D. Krasner, ed., *International Regimes*, 337–354. Ithaca, NY: Cornell University Press.

Stumme, Gerd. 2000. *Conceptual On-Line Analytical Processing*. Dordrecht: Kluwer Academic Publishers.

Stumme, Gerd, und Rudolf Wille, eds. 2000. *Begriffliche Wissensverarbeitung—Methoden und Anwendungen*. Berlin: Springer.

Take, Ingo. 2002. *NGOs im Wandel: Von der Graswurzel auf das diplomatische Parkett*. Wiesbaden: Westdeutscher Verlag.

Tetlock, Phillip E., and Aaron Belkin. 1996. "Counterfactual Thought Experiments in World Politics: Logical, Methodological, and Psychological Perspectives." In Phillip E. Tetlock and Aaron Belkin, eds., *Counterfactual Thought Experiments in World Politics: Logical, Methodological, and Psychological Perspectives*, 3–38. Princeton, NJ: Princeton University Press.

Tyler, Tom R. 1990. *Why People Obey the Law*. New Haven: Yale University Press.

Underdal, Arild. 1998. "Explaining Compliance and Defection." *European Journal of International Relations* 4(1): 5–30.

Underdal, Arild. 2002. "One Question, Two Answers." In Edward L. Miles, Arild Underdal, Steinar Andresen, Jørgen Wettestad, Jon Birger Skjaerseth, and Elaine M. Carlin, *Regime Effectiveness: Confronting Theory with Evidence*, 1–45. Cambridge, MA: MIT Press.

Underdal, Arild, and Oran R. Young, eds. 2004. *Regime Consequences: Methodological Challenges and Research Strategies*. Dordrecht: Kluwer Academic Publishers.

Victor, David G., Kal Raustiala, and Eugene B. Skolnikoff, eds. 1998. *The Implementation and Effectiveness of International Environmental Commitments: Theory and Practice*. Cambridge, MA: MIT Press.

Vogel, David, and Timothy Kessler. 1998. "How Compliance Happens and Doesn't Happen Domestically." In Edith Brown Weiss and Harold K. Jacobson, eds., *Engaging Countries: Strengthening Compliance with International Environmental Accords*, 19–37. Cambridge, MA: MIT Press.

Vogt, F., C. Wachter, and R. Wille. 1991. "Data Analysis Based on a Conceptual File." In H. H. Brock and P. Ihm, eds., *Classification, Data Analysis and Knowledge Organization*, 131–142. Berlin: Springer.

Waltz, Kenneth. 1979. *Theory of International Relations*. New York: Random House.

Weaver, R. Kent, and Bert A. Rockman, eds. 1993. *Do Institutions Matter? Government Capabilities in the United States and Abroad*. Washington, DC: Brookings Institution.

Weiler, Joseph H. H. 1993. "Journey to an Unknown Destination: A Retrospective of the European Court of Justice in the Arena of Political Integration." *Journal of Common Market Studies* 31(4): 417–446.

Weiss, Edith Brown, and Harold K. Jacobson, eds. 1998. *Engaging Countries: Strengthening Compliance with International Environmental Accords*. Cambridge, MA: MIT Press.

Weiss, Edith Brown, Paul C. Szasz, and Daniel B. Magraw. 1992. *International Environmental Law: Basic Instruments and References*. Ardsley, NY: Transnational Publishers.

Wendt, Alexander. 1999. *Social Theory of International Politics*. Cambridge: Cambridge University Press.

Werksmann, Jacob, ed. 1996. *Greening International Institutions*. London: Earthscan.

Wettestad, Jørgen. 1999. *Designing Effective Environmental Regimes: The Key Conditions*. Cheltenham: Edward Elgar.

Wettestad, Jørgen. 2002. *Clearing the Air—European Advances in Tackling Acid Rain and Atmospheric Pollution*. Aldershot: Ashgate.

Whitehorn, Mark, and Bill Marklyn. 2001. *Inside Relational Databases*. 2nd ed. London: Springer.

Wille, R. 1992. "Begriffliche Datensysteme als Werkzeug de Wissenskommunikation." In H. H. Zimmermann, H. D. Luckwalt, and A. Schulz, eds., *Mensch und Maschine—Informationelle Schnittstellen der Kommunikation*, 63–73. Konstanz: Universitätsverlag Konstanz.

Wolf, Dieter. 2005. "State Aid Control at the National, European and International Level." In Michael Zürn and Christian Joerges, eds., *Law and Governance in Postnational Europe: Compliance beyond the Nation-State*. Cambridge: Cambridge University Press.

Young, Oran R. 1979. *Compliance and Public Authority: A Theory with International Application*. Baltimore: Johns Hopkins University Press.

Young, Oran R. 1994. *International Governance: Protecting the Environment in a Stateless Society*. Ithaca, NY: Cornell University Press.

Young, Oran R., ed. 1999a. *The Effectiveness of International Regimes: Causal Connections and Behavioral Mechanisms*. Cambridge, MA: MIT Press.

Young, Oran R. 1999b. *Governance in World Affairs*. Ithaca, NY: Cornell University Press.

Young, Oran R. 2002a. "Are Institutions Intervening Variables or Basic Causal Forces? Causal Clusters versus Causal Chains in International Society." In Michael Brecher and Frank Harvey, eds., *Millennium Reflections on International Studies*, 176–191. Ann Arbor: University of Michigan Press.

Young, Oran R. 2002b. "Evaluating the Success of International Environmental Regimes: Where Are We Now?" *Global Environmental Change* 12: 73–77.

Young, Oran R. 2002c. *The Institutional Dimensions of Environmental Change: Fit, Interplay, and Scale*. Cambridge, MA: MIT Press.

Young, Oran R., with contributions from Arun Agrawal, Leslie A. King, Peter H. Sand, Arild Underdal, and Merrilyn Wasson. 1999. *Institutional Dimensions of Global Environmental Change (IDGEC) Science Plan.* IHDP Report No. 9. Bonn: IHDP.

Young, Oran R., and Gail Osherenko, eds. 1993. *Polar Politics: Creating International Environmental Regimes*. Ithaca, NY: Cornell University Press.

Zangl, Bernhard. 1999a. *Interessen auf zwei Ebenen. Internationale Regime in der Agrarhandels, Währungs- und Walfangpolitik*. Baden-Baden: Nomos.

Zangl, Bernhard. 1999b. *Internationale Normdurchsetzung: Enforcement, Management oder Adjudication?* Bremen: InIIS Arbeitspapier 15/99.

Zürn, Michael. 1992. *Interessen und Institutionen in der internationalen Politik: Grundlegung und Anwendung des situationsstrukturellen Ansatzes*. Opladen: Leske und Budrich.

Zürn, Michael. 2004. "Global Governance and Legitimacy Problems." *Government and Opposition* 39(2): 260–287.

Zürn, Michael, and Christian Joerges, eds. 2005. *Law and Governance in Postnational Europe: Compliance beyond the Nation-State*. Cambridge: Cambridge University Press.

Zürn, Michael, and Dieter Wolf. 1999. "European Law and International Regimes: The Features of Law beyond the Nation State." *European Law Journal* 5(3): 272–292.

Zürn, Michael, and Jürgen Neyer. 2005. "Conclusions: The Conditions of Compliance." In Michael Zürn and Christian Joerges, eds., *Law and Governance in Postnational Europe: Compliance beyond the Nation-State*. Cambridge: Cambridge University Press, 183–217.

# Index

**Global Environmental Accord: Strategies for Sustainability and Institutional Innovation**

Nazli Choucri, series editor

*(continued from page ii)*

Corey L. Lofdahl, *Environmental Impacts of Globalization and Trade: A Systems Study*

Oran R. Young, *The Institutional Dimensions of Environmental Change: Fit, Interplay, and Scale*

Tamar L. Gutner, *Banking on the Environment: Multilateral Development Banks and Their Environmental Performance in Central and Eastern Europe*

Liliana B. Andonova, *Transnational Politics of the Environment: The European Union and Environmental Policy in Central and Eastern Europe*

David L. Levy and Peter J. Newell, editors, *The Business of Global Environmental Governance*

Dennis Pirages and Ken Cousins, editors, *From Resource Scarcity to Ecological Security: Exploring New Limits to Growth*

Ken Conca, *Governing Water: Contentious Transnational Politics and Global Institution Building*

Sebastian Oberthür and Thomas Gehring, editors, *Institutional Interaction in Global Environmental Governance: Synergy and Conflict Among International and EU Policies*

Ronald B. Mitchell, William C. Clark, David W. Cash, and Nancy M. Dickson, editors, *Global Environmental Assessments: Information and Influence*

Helmut Breitmeier, Oran R. Young, and Michael Zürn, *Analyzing International Environmental Regimes: From Case Study to Database*